A NATURALIST'S GUIDE TO THE

MAMMALS
OF
SOUTHEAST ASIA

Brunei, Cambodia, Indonesia, Laos,
Malaysia, Myanmar, the Philippines,
Singapore, Thailand and Vietnam

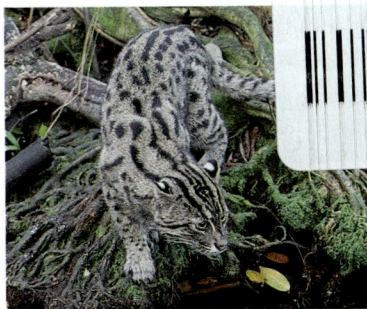

Chris R. Shepherd and Loretta Ann Shepherd
3rd edition revised by Charles M. Francis

jb

JOHN BEAUFOY PUBLISHING

This edition published in the United Kingdom in 2024 by John Beaufoy Publishing
11 Blenheim Court, 316 Woodstock Road, Oxford OX2 7NS, England
www.johnbeaufoy.com

10 9 8 7 6 5 4 3 2 1

ISBN 9781913679682

Photo captions
Front cover: *main image* Asian Elephant (Shutterstock/Mike Workman); *bottom left* Tiger (Stephen Hogg/
Wildtrack Photography; *bottom centre* Sun Bear (Wong Siew Te/Bornean Sun Bear Conservation Centre); *bottom
right* Cream-coloured Giant Squirrel (Nick Baker, EcologyAsia.com). **Back cover:** Dusky Langur (Chris R.
Shepherd). **Title page:** Fishing Cat (Abraham Matthew/Singapore Zoo and Night Safari). **Contents page:** Siamang
(Vilma D'Rozario/Cicada Tree Eco-Place).

Illustrations by Stephen Dew

Dedication
For our daughter Raven Dhanya Shepherd.

Edited and designed by D & N Publishing, Baydon, Wiltshire, UK

Printed and bound in Malaysia by Times Offset (M) Sdn. Bhd.

·Contents·

INTRODUCTION

Globally, there are over 6,600 species of mammals. At least 950 occur naturally in Southeast Asia, and many of these are found nowhere else in the world. An increasing number of Southeast Asia's mammals are severely threatened – directly by hunting and the wildlife trade, and indirectly by habitat loss, urbanisation, the introduction of non-native species, etc. Conservation actions for the region's mammals have never been so urgently needed.

Despite the region boasting such an amazing wealth of species, and despite many of these slipping precariously close to the edge of extinction, most people have very limited knowledge, awareness or appreciation of these animals or their needs. Fewer still realise that each of us can play a role in ensuring that this incredibly varied yet intertwined myriad of species remains intact for generations to come.

This book is intended to play a part in just that – raising awareness and interest in the region's mammals, and encouraging more people to get involved in protecting these species and their fragile habitats. We sincerely hope that you enjoy the photographs featured here and find the information useful. We hope that with this guide you will have a greater appreciation for Southeast Asia's wildlife, and that you will be inspired to play your part, be it by becoming a more responsible citizen, by reducing your impact on the environment around you, by pledging to support individuals or organisations working to better understand and protect mammals and other wildlife, or by choosing to become directly involved in research and conservation yourself.

Southeast Asia here refers to the countries of Brunei, Cambodia, Indonesia (Greater Sundas only – Java, Borneo, Sumatra and associated smaller islands, with the line being drawn immediately east of Bali), Laos, Malaysia, Myanmar, the Philippines, Singapore, Thailand and Vietnam. This area is home to a vast variety of mammals living in an amazingly wide range of habitats, from mountaintops to the sea, brackish mangroves and vast wetlands, to tall rainforests, dry open forests and rugged limestone hills. Increasingly, secondary forest habitats are home to many species and are therefore worthy of conservation efforts as well. Some more adaptable species are even found in plantations and agricultural areas, and living alongside humans in rural and occasionally even urban settings – though it is important to recognise that the more habitat is lost, the more the variety and viability of the mammal communities living within suffers.

Riverine species, such as the Proboscis Monkey (*Nasalis larvatus*), can be viewed fairly easily on the Kinabatangan river in Sabah, Malaysia.

USING THIS BOOK

This book is not intended to be a comprehensive field guide, instead aiming to introduce readers to a selection of the 950 species of mammals native to Southeast Asia. There are 129 species accounts featured in these pages, with photographs and notes on each, including a basic description, information on the species' range and preferred habitats, and a variety of other interesting points.

The majority of the images we have used are of wild mammals, photographed in their natural state, but a few are of captive individuals, largely because of the rarity of the species in question, or because of the unavailability of suitable wild images showing the identification features. A number are remote camera-trap images taken by researchers. These camera traps, as they are usually referred to, have become essential tools in the field of mammal research and conservation. With such devices positioned in the forest, invaluable photographs and information can be collected – anything from basic presence or absence to far more detailed data, such as aiding in identifying individual Tigers (*Panthera tigris*) by their unique stripe patterns. Camera traps can remain far longer in the field than a researcher, and are much less disruptive to the wildlife. We have deliberately included a few pictures taken with camera traps, although the picture quality is sometimes lower than that of a photo taken in captivity, to give the reader a sense of the research being carried out.

A brief description of the main features of each species is given, to assist in identification. Standard measurements are given as follows (see also diagrams below):

HB   head and body
T   tail length
TL   total length (for cetaceans and sirenians)

SH   shoulder height (for larger mammals)
FA   forearm (for bats)

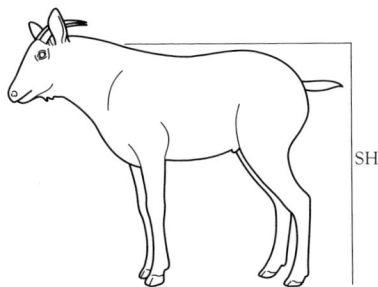

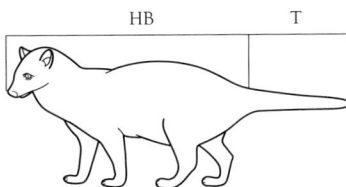

Measurements refer to adult animals. Remember that sizes may vary greatly, and in some cases there is great variety in both size and colour within a single species.

A brief description of the known habitat, or habitats, used by each species is also given. A few notes are also provided on the habits or ecology of the species. It is important to remember that habitat use of many species may vary, sometimes seasonally, sometimes due to the loss of preferred habitat. Others are generalists and use a wide variety of habitats or are able to adapt not only to a number of natural settings but also to alterations made by humans. Some species can adapt to extreme habitat alteration, while others leave or simply perish once their habitat has been disturbed. Loss of habitat is one of the greatest threats to Southeast Asia's mammals.

We have included a checklist of the mammals found in the region, which we have attempted to make as complete and accurate as possible, based on available information – though we acknowledge that there may be omissions. We recognize this will continue to change, as new mammals are being discovered all the time. We have relied heavily on the IUCN Red List of Threatened Species, and various regional field guides and country species lists (see Further Information), especially *Field Guide to the Mammals of South-East Asia* by Charles M. Francis, as well as the Mammal Diversity Database maintained by the American Society of Mammalogists. Introduced species, and species already considered extinct, such as Schomburgk's Deer (*Rucervus schomburgki*), are not included in this list. Others, such as the Kouprey (*Bos sauveli*), which is likely to be extinct, and the Sika (*Rucervus nippon*), which is likely to be extirpated from Southeast Asia, are still included in the checklist, in the countries they were originally found in, as they have not yet been officially declared extinct or extirpated. Included also is the current status of each species according to the IUCN Red List. Since the status of species is periodically revised, readers should check the current status at www.iucnredlist.org.

WHAT'S IN A NAME?

All species have a scientific (usually 'Latinised') name. This nomenclature is to avoid confusion, as common names vary in different languages and locations, but the scientific binominal (two-part) names are used universally. The first part of the scientific name – *Panthera*, for example – denotes the genus, the second part – e.g. *tigris* – the species within that genus: *Panthera tigris*, the Tiger. A third name is added when referring to a subspecies: *Panthera tigris sumatrae*, the Sumatran Tiger.

Common names used, in English, are also given in this book, to make it convenient for people unfamiliar with scientific names. However, common names can be confusing, as there is often more than one for a single species (for example, Silvered Langur, Silvered Leaf Monkey, Silvered Monkey, Silvery Lutung are all names for *Trachypithecus cristatus*). In an attempt to standardise the common names in the region, where possible we have used the same names as Charles M. Francis in *A Field Guide to the Mammals of South-East Asia*. As this is the most recent and comprehensive guide for the mainland of Southeast Asia, it makes little sense to vary the names yet again and further confuse the issue. For example, we have taken the lead from Francis and eliminated the terms leaf monkey, surili and lutung, instead calling them all langurs. Having said that, some of the names in this book do differ or are slightly modified, based on recommendations from Duckworth and Pine's 2003 paper on 'English names for a world list of mammals' (see Further Information). Furthermore, the geographic coverage in Francis's book does not include Brunei, Indonesia, Malaysian Borneo (Sabah and Sarawak), the Philippines and Singapore, and therefore names were obtained elsewhere for species occurring in these countries.

The taxonomy of the region's mammals is far from being finalised, and much more work in this field is urgently needed. Many of the species are likely in fact to represent complexes of more than one species. What now appears to be a single widespread and common species may actually include a number of localised and threatened species that require urgent conservation assistance.

OPPORTUNITIES FOR NATURALISTS

While Southeast Asia is one of the wealthiest regions in the world in terms of species diversity, very little is actually known about the ecology, habits, status, threats and conservation needs of the majority of its mammals. It is hoped that with more people – biologists, naturalists, local inhabitants and visitors from overseas – observing and learning about mammals in their natural habitat, the result will be that these information gaps are gradually filled.

Montane forests are important habitats for a variety of species throughout the region.

MAMMAL WATCHING

While many amazing mammals occur here, finding them is the trick. In some parts of the region, some are quite easy to find, especially diurnal mammals that can live in close proximity to people (in areas where people are not hunting them), such as some monkeys and squirrels. But the vast majority of them are more challenging. Some mammal species have only been seen a few times ever!

Some simple tools are a good pair of binoculars, identification guides and a notebook – and a torch or headlamp to find nocturnal mammals. A healthy dose of patience is also useful. Finding mammals is not always easy, and once they have been detected, they often flee. Many are nocturnal and/or arboreal, further adding to the list of challenges. It is important to follow some basic principles of keeping noise levels low, wearing mute-coloured field clothes, moving quietly and cautiously, and paying attention to animal sounds and signs. The call of gibbons, for instance, may reveal a dueting pair, marks on trees may be clues to the presence of bears, and busy footprints may lead to a herd of deer.

Identifying mammals – especially small mammals, where clear views are not always possible because they are hidden or are fleeing, and which in many cases closely resemble other species – takes some practice. Do not be discouraged. Take notes on what you have seen, including description, location, habitat, altitude, behaviour, and other details of the encounter. This will help confirm the identity of the mystery mammal, and could very likely contribute to the current pool of knowledge.

Very little is known about most of the mammals of Southeast Asia. Increased knowledge, whether through basic field observations or from intense research, is desperately needed in order better to understand the needs of each and ultimately to try and prevent the loss of any more species from the wild. Acquiring suitable photographs for this book was challenging, highlighting the need for more photographers, amateur and professional, to get out there and take pictures, especially of the more secretive and little-known species. There is not a great deal of information on most of the region's mammals, so every observation, record and photograph taken is potentially an important contribution to the overall understanding of that species' behaviour, ecology and requirements.

WILDLIFE IN TROUBLE

Sadly, an increasing number of Southeast Asia's mammals are threatened. At the time of writing this book more than 190 are considered Threatened (94 Vulnerable, 75 Endangered, 21 Critically Endangered) in the IUCN Red List, the most comprehensive evaluation of the conservation status of the world's plant and animal species (www.iucnredlist.org). Human-related activities are directly or indirectly responsible for the threatened status, and more species are being added to the list all the time.

HABITAT LOSS

Habitat loss and fragmentation is a serious threat to the conservation of most mammals in Southeast Asia. Development – residential and agricultural expansion, especially for oil palm

Habitat loss: a severe threat to many of the region's mammals.

– has replaced large areas of habitat crucial to many mammal species. Highways and roads cut through forests, not only reducing the available space and limiting the movements of many species, but also increasing access for hunters. Logging activities degrade and destroy pristine habitat. While a few mammal species can survive in these disturbed habitats, and a few more at the edges, most cannot. Large-scale monoculture plantations in particular are a major threat to the continued survival of many species, destroying habitats, isolating populations, and bringing some species, such as Tigers, Leopards (*Panthera pardus*), Asian Elephants (*Elephas maximus*), Asian Black Bears (*Ursus thibetanus*) and Eurasian Wild Pigs (*Sus scrofa*), into conflict with people. In almost all such conflicts, the wildlife is the loser.

Steps must be taken to minimise the negative impact of development. Key habitats and connecting corridors must be set aside. Forests along rivers and other water bodies should be left intact, to maintain habitat and to allow wildlife access to water. Long-term land-use planning, be it for plantations, roads, farming or other development, must take the conservation of wildlife and wild places into consideration. In Southeast Asia, where human population growth and expansion is taking place at a frightening pace, this is a major challenge. But not impossible.

TRADE IN WILDLIFE

Illegal and unsustainable trade, both domestic and international, is a growing threat to a rapidly increasing number of species, and nowhere is this more obvious and urgent than in Southeast Asia. Wildlife has been harvested and traded throughout this region for thousands of years, yet never have the levels of harvesting and trade been as intense and as destructive as has been observed over the past few decades. Amongst all the threats wildlife faces, illegal trade is an extremely urgent issue that needs the highest level of attention, as it has the greatest potential to do maximum harm in a short time.

Humans have always exploited wildlife. Since the early times, our ancestors relied on meat from wild animals, and used their skins for clothing and shelter. As early as 2600 BC, rhinoceros horn and other animal parts were used in medicines. Emperors and other rulers kept elephants, to be used in war and ceremonies, and elephant ivory has been traded for centuries. There are records, for example, of Thailand exporting ivory to China and Japan dating back several centuries. In the 13th century, Thailand exported tusks to Fukien Province, China, and in the 19th century there is a record of Thailand exporting 18 tonnes of tusks to China.

Today, many of Southeast Asia's mammals are severely threatened by trade – and in many cases this poses a greater threat than habitat loss. Civets are kept in cramped cages, fed coffee beans for the civet coffee or coffee luwak industry. Primates are eaten, as well as being traded for biomedical research and the pet industry. Cats are traded for their bones, meat and skins, while bears are hunted for their paws and gallbladders, or captured and kept in bear bile extraction facilities.

Many of these species, like the elephants killed to supply ivory demand centuries ago, are hunted to satisfy the demand from East Asia, especially China. Tigers and rhinos are prime examples, with their bones and horns, respectively, being highly prized in traditional Asian medicine. This demand has caused these species to become all but extinct in the wild. Suitable habitat for both Tigers and rhinos still remains in many places, but the animals are gone.

More resources and efforts must be channelled towards preventing any more species being lost to these threats. It is, however, not too late. We hope this book will encourage you to do what you can to support the conservation of Southeast Asia's mammals.

GLOSSARY

agouti – coat colour type that has banding or stripes on individual hairs

allomothering – non-maternal care of infants

anthropogenic – caused by humans

antlers – bony, branched structures protruding from the front of the skull, shed annually, unlike horns, which are not branched and are permanent

arboreal – tree-dwelling

axillary – near the armpit region

beak – dolphin's snout

bovid – member of the Bovidae (cattle) family

brachiate – to move by swinging on the arms from one handhold to another

cephalopod – marine molluscs such as cuttlefish, squid and octopus

cetacean – whales, dolphins and porpoises

colobine – member of the Colobinae subfamily of primates

commensal – two species living in close association without harming each other

crepuscular – active at dawn and dusk

dew toes – hind toes that are a little smaller than the hoofs, which do not usually reach the ground except when there is soft soil

dipterocarp – tall, hardwood tropical tree species in the family Dipterocarpaceae that can grow to an exceptionally large size

distal – away from the body, towards the end

diurnal – active during the day

dorsal – the back surface or back part of the body

echolocation – determining the location of an object by relying on echoes of sounds emitted

endemic – native species restricted to a particular geographic area

epiphytic – a plant that grows attached to another plant but does not cause the host any damage

extirpated – species no longer exists in a country or region

falcate – sickle-shaped curve

family, genus and species – scientific classifications are used to categorise organisms into seven major divisions, which are known as taxa. The final three (family, genus and species) are the most specific of the categories, and show how closely related organisms are. For instance, members of the same genus are more closely related than members of the same family

fluke – each lobe of the tail of a cetacean or sirenian

frugivorous – having a diet mainly of fruit

fusiform – tapering at both ends

genus – *see* family, genus and species

horn – paired, bony, permanent growth on the upper part of the heads of certain hoofed mammals such as cattle, sheep and goats, usually curved or pointed; also similar pointed growth on the snout of rhinoceros

interfemoral membrane – the skin between the hind legs and tail of a bat, also known as the uropatagium

karst – rough and rocky landscape that comprises caves, underground channels and sinkholes, which is formed when the underground water dissolves soluble layers of bedrock such as limestone and dolomite

melanistic – high concentration of dark pigmentation that causes an overall appearance of dark colouration

melon – large and rounded forehead area of a dolphin

montane – mountainous area

morph – form or type within a species, usually referring to a colour variation with no taxonomic significance

morphology – physical form and structure of an organism

nocturnal – active during the night

noseleaf (bats) – skin around the nose

pantropical – throughout the tropics

patagium – gliding membrane between fore and hind limbs, enclosing the tail

pedicel (deer) – bony base on forehead that supports the antlers

pelage – coat of a mammal (such as fur, wool, hair)

prehensile – tip (mainly of tail, snout, lips) that can curl and grasp objects

premaxilla – small bones at the tip of the jaw, supporting incisors

primary habitat/forest – old-growth forest that has never been logged

proboscis – long and flexible snout

sacculated – stomach with many chambers that aids digestion of high volumes of plant material

secondary habitat/forest – forest that has been logged but has recovered

species – *see* family, genus and species

split (taxonomy) – dividing further, to represent more than one group

subspecies – division within a species, distinct from other members of the same species, but not so distinct as to be considered a separate species

sub-montane – foothills of mountainous areas

sympatric – more than one species sharing the same geographic area

taxonomy – scientific system of classifying and identifying organisms

terrestrial – ground-dwelling

tine – prong on an antler

tusk (elephants, dugongs, etc.) – elongated, protruding tooth, usually in pairs

ventral – underside

vibrissae – long, stiff whisker hairs on snout and brows

PANGOLINS
Also known as scaly anteaters, pangolins were once grouped together with anteaters, sloths, armadillos and the Aardvark (*Orycteropus afer*) – but research confirms that these species are not closely related although they have similar adaptations.

Sunda Pangolin ◾ *Manis javanica* HB 40–65cm, T 35–56cm

DESCRIPTION The entire upperparts, including the tail, are covered in brownish scales. It has very small ears, and a long and tapered head. Its underparts have no scales. It has very long claws.

DISTRIBUTION Brunei, Cambodia, Indonesia (Java, Kalimantan, Sumatra), Laos, Malaysia (Peninsular Malaysia, Sabah and Sarawak), Myanmar, Singapore, Thailand and Vietnam.

HABITS AND HABITAT Largely nocturnal, sleeping in an underground burrow during the day. Tall and secondary forest are the preferred habitat. Highly specialised diet of ants and termites, which it extracts using its strong claws and long sticky tongue from nests in trees, on and below the ground. The prehensile tail is especially useful when the pangolin is climbing trees.

NOTES When alarmed, it curls into a ball and wraps its tail around itself to protect its non-scaly underparts. It is heavily poached and traded for its purported medicinal properties, with populations declining severely throughout its range.

Chinese Pangolin ■ *Manis pentadactyla* HB 40–58cm, T 25–38cm

DESCRIPTION The upperparts, including the tail, are covered in brown scales, and it has more prominent ears than the Sunda Pangolin (see p. 12). The scales on its head extend only part-way to the nostrils. It has long claws, which are longer on the forelimbs. Its tail is less flexible than the Sunda Pangolin's and considerably shorter, relative to the head-and-body length.

DISTRIBUTION Laos, Myanmar, Thailand and Vietnam. Also found in Bangladesh, Bhutan, China, Hong Kong, India, Nepal and Taiwan.

HABITS AND HABITAT The Chinese Pangolin is very similar to the Sunda Pangolin, also being nocturnal, living in tall primary and secondary forest and having a highly specialised diet of termites and ants.

NOTES This species has been seriously impacted by poaching and trade for its purported medicinal properties.

> ### Moonrats and Gymnures
> Moonrats and gymnures belong to the Erinaceidae family, which also includes hedgehogs, but they lack sharp spines.

Moonrat ◾

Echinosorex gymnura
HB 32–40cm, T 20–29cm

DESCRIPTION In most of the range the front part of the body is white to greyish white and the remainder is black with greyish frosting. The tail is long and scaly, with short hair, dark at the base and light towards the end. It has an elongated snout with a pinkish nose. In Borneo, and occasionally elsewhere, this large insectivore is all white with a sparse scattering of black hairs.

DISTRIBUTION Brunei, Indonesia (Kalimantan and Sumatra), Malaysia (Peninsular Malaysia, Sabah and Sarawak), southernmost Myanmar and southernmost Thailand.

HABITS AND HABITAT Primary and secondary lowland forests, including mangroves and swamp forests. Also found in hill forests. The species can also tolerate some habitat disturbance. From sea level to 1,000m.

NOTES The moonrat has a strong pungent odour.

Short-tailed Gymnure ◾

Hylomys suillus HB 12–14cm, T 2–3cm

DESCRIPTION Upperparts reddish brown to dark brown, with a grey tinge. Underparts light grey, with white-tipped hairs. Resembles a shrew, with long snout, but has a very short hairless tail.

DISTRIBUTION Brunei, Cambodia, Indonesia (Java, Kalimantan, Sumatra), Laos, Malaysia (Peninsular Malaysia, Sabah and Sarawak), Myanmar, Thailand and Vietnam. Also in China.

HABITS AND HABITAT Active day and night. In hill and montane forests up to 3,000m, but sometimes in humid lowland forests. Feeds mainly on insects but also takes some fruit.

NOTES Recent research indicates this is actually a complex of at least seven different species separated geographically (see checklist p. 154)

Mountain Treeshrew ■ *Tupaia montana* HB 15–22cm, T 13–19cm

DESCRIPTION The upperparts of this species vary from dark brown to reddish or olive-brown, always with fine reddish speckling. Its underparts are buffy red. Some individuals have a pale shoulder stripe. Very similar to the Common Treeshrew (see p. 16), but has smaller hind feet and usually occurs at higher altitudes.

DISTRIBUTION Endemic to Borneo: Brunei, Indonesia (Kalimantan) and Malaysia (Sabah and Sarawak).

HABITS AND HABITAT Diurnal, though most active in early morning and late afternoon. Found only in sub-montane and montane primary and slightly disturbed secondary forests, usually above 600m. Largely terrestrial.

NOTES This is one of the most easily observed mammals in montane areas of Sabah.

Common Treeshrew ■ *Tupaia glis* HB 13–21cm, T 12–20cm

DESCRIPTION Reddish-brown body with darker brownish-grey tail and head. Underside lighter to yellowish, often with a whitish stripe over the shoulder. Light eye-ring. Long pointed muzzle with sharp pointed teeth. It has short limbs and a long squirrel-like bushy tail.

DISTRIBUTION Malaysia (Peninsular Malaysia), Singapore and Thailand.

HABITS AND HABITAT Diurnal. Mainly in lowland forests below 1,500m. Also in parks and plantations, and in some places, such as in Peninsular Malaysia, has adapted to suburban gardens. These small mammals feed on insects and fruit. Usually seen alone or in pairs, foraging on the ground or in low trees and bushes.

NOTES The name 'treeshrew' is misleading, as they are not shrews. Looking more like squirrels than shrews, the treeshrews have long been a family of confused identity – they were once even considered primates but are now placed in their own order.

COLUGOS

Colugos are also known as flying lemurs, but this is an obvious misnomer as they glide rather than fly, and they aren't lemurs. They have a gliding membrane called the patagium between fore and hind limbs, enclosing the tail. There are a number of other mammals that glide from tree to tree, such as flying squirrels (pp. 126–130), but their feet and tail are free of the patagium.

Sunda Colugo ▪ *Galeopterus variegatus* HB 33–42cm, T 18–27cm

DESCRIPTION The Sunda Colugo is grey to reddish brown. The shaded and mottled coat camouflages the animal perfectly against tree bark.

DISTRIBUTION Brunei, Cambodia, Indonesia (Java, Kalimantan, Sumatra), Laos, Malaysia, Myanmar, Thailand, Singapore and Vietnam.

HABITS AND HABITAT Nocturnal but sometimes active during the day. Found in evergreen forest below 1,000m above sea level. Completely arboreal, clinging to the sides of trees and gliding between tall trees. Diet mainly leaves and flowers.

NOTES Females carry their young enclosed in the patagium. Some males have also been observed doing the same, though it is unlikely that they play a very large parenting role.

LEFT: *Adult.* RIGHT: *Juvenile*

Philippine Colugo ▪ *Cynocephalus volans* HB 33–38cm, T 17–27cm

DESCRIPTION Brown or grey-brown dorsal fur, with the males darker than the females. Has a patagium like the Sunda Colugo (see p. 17).

DISTRIBUTION Philippines.

HABITS AND HABITAT Found in lowland primary, secondary and mixed forest and also disturbed habitat. Diet mainly leaves, rarely fruit. Nocturnal and crepuscular, completely arboreal, also clinging to the sides of trees and gliding between tall trees.

Large Flying-fox ■ *Pteropus vampyrus* HB 27–32cm, T absent, FA 18–20cm

DESCRIPTION Most fruit bats rely on sight and smell to find their way, which is why they have large, reflective eyes. The Large Flying-fox has one of the greatest wingspans of any bat, reaching 1.5m. Fur colour is variable, but usually the body is black while the upper shoulders, the chest and the back of the head are russet or orange-brown. The sides of the head are reddish brown, blending into the black underparts. Juveniles are uniform grey-brown in colour. The wings in both adults and juveniles are dark brown.

DISTRIBUTION Distribution: Brunei, southern Cambodia, Indonesia (Bali, Java, Kalimantan, Sumatra), Malaysia (Peninsular Malaysia, Sabah, Sarawak), southern Myanmar, Philippines, Singapore, Thailand and southern Vietnam. Also found in Timor-Leste.

HABITS AND HABITAT Found in primary and secondary forest, often in mangroves and nipah palm. While sometimes found in disturbed forests, they prefer undisturbed and riparian forest. Found up to at least 1,250m. These giant bats roost high up in large trees in colonies from a few dozen up to formerly nearly 100,000 individuals. Unfortunately, in many parts of its range populations have been greatly reduced, or completely eliminated, largely due to hunting and the species is now considered Endangered.

NOTES The Large Flying-fox feeds on a wide range of wild and cultivated fruits (especially figs), flowers and leaves, and it is an extremely important pollinator for many trees that are valuable to local communities and for commerce, including durian.

Spotted-winged Fruit Bat ■ *Balionycteris maculata*

HB 5.3–6.2cm, T absent, FA 4–4.5cm

DESCRIPTION One of the smallest fruit bats in Asia. Dark blackish brown, with an even darker head. The underparts are grey, and it lacks an external tail. Its wing membrane is dark brown, flecked with small whitish spots, especially on the joints, and it has pale spots on the muzzle of its dog-like face, just in front of each eye. The nostrils are elongated, almost tube-like.

DISTRIBUTION Brunei, Indonesia (Kalimantan, Sumatra), Malaysia (Sabah, Sarawak). Populations in Peninsular Malaysia and Thailand are now considered a separate species, *B. seimundi*.

HABITS AND HABITAT Found in primary lowland forests, and occasionally montane forests, from sea level up to 1,500m. In some parts of its range it has been recorded from secondary forests. It roosts in small groups in the crowns of palms, cavities in epiphytic plants, tree-hollows, and cavities in arboreal insect nests. Gives birth to a single young up to twice a year.

NOTES The Spotted-winged Fruit Bat is threatened throughout its range by habitat loss.

ABOVE: *Four Spotted-winged Fruit Bats roosting together*

Lesser Short-nosed Fruit Bat ▪ *Cynopterus brachyotis*

HB 7–8cm, T 0.8–1cm, FA 6–7cm

DESCRIPTION A moderately small fruit bat. Brown to yellowish brown with an orange collar. The collar is more yellowish in colour in females. Pale edges to the ears, and whitish wing bones visible through translucent skin. Has a broad snout.

DISTRIBUTION Brunei, Cambodia, Indonesia (Java, Kalimantan, Sumatra), Laos, Malaysia (Peninsular Malaysia, Sabah, Sarawak), Myanmar, Philippines, Singapore, Thailand and Vietnam. Presence in Brunei uncertain. Also found in China, India, Sri Lanka and Timor-Leste.

HABITS AND HABITAT Lowland primary and secondary forests, mangroves, orchards, parks and gardens. It roosts in palms, under shaded trees, or near cave entrances in rural and urban landscapes and in forested areas, either solitary or in small groups. It feeds on small fruits, figs and nectar.

NOTES Several similar-looking species of short-nosed fruit bats occur in the region differing mainly in size. They are important dispersers of many secondary forest fruits.

Cave Nectar Bat ▪ *Eonycteris spelaea* HB 8.5–11cm, T 1.5–1.8cm, FA 6–7cm

DESCRIPTION Upperparts grey-brown to dark brown. Underparts paler. Neck sometimes yellowish brown. The muzzle is elongated, adapted for drinking nectar. Short external tail.

DISTRIBUTION Brunei, Cambodia, Indonesia (Java, Kalimantan, Sumatra), Laos, Malaysia (Peninsular Malaysia, Sabah, Sarawak), Myanmar, Philippines, Singapore, Thailand and Vietnam. Also found in China, India and Timor-Leste.

HABITS AND HABITAT Found in primary forests and in disturbed and agricultural areas. Roosts in caves, in large groups, with some roosts exceeding 50,000 individuals. Sometimes roosts with other bat species. In some places, this species seems to have adapted well to leafy, semi-urban habitats. Travels many kilometres each night in search of the nectar of flowering trees and shrubs.

NOTES This species is an important pollinator of fruit trees, such as durian.

Malayan Slit-faced Bat ■ *Nycteris tragata*

HB 6.5–7.5cm, T 6.5–8cm, FA 4.5–5.5cm

DESCRIPTION Fur orange-brown to grey-brown, slightly paler below. Long oval ears with a short tragus. Nose with a series of skin flaps on either side of a narrow slit. Very long tail with a forked tip.

DISTRIBUTION Brunei, Indonesia (Borneo, Sumatra), Malaysia (Peninsular Malaysia, Sabah, Sarawak), southern Myanmar and Thailand.

HABITS AND HABITAT Found mainly in mature forest in the lowlands, though will use disturbed forests with tall trees. Roosts in small groups in hollow trees or in caves. Feeds mainly on large insects that it gleans from surfaces such as branches or the ground; may hunt by listening for the sounds they make.

NOTES A closely related species, *N. javanicus*, occurs in Java. The remaining 13 species in the family are found only in Africa.

Greater False-vampire ▪ *Lyroderma lyra*
HB 6.5–9.5cm, T absent, FA 6.5–7.2cm

DESCRIPTION Upperparts greyish brown with long fur. Underparts paler. No visible tail. Very large ears, joined at the base. Posterior lobe of noseleaf elongate with stiffened central ridge, approximately parallel-sided flaps on the sides and squared off at the top. Intermediate noseleaf is narrower than the anterior noseleaf. Anterior noseleaf does not cover the protruding muzzle.

DISTRIBUTION Cambodia, Laos, Malaysia (Peninsular Malaysia), Myanmar, Thailand and Vietnam. Also found in Afghanistan, Bangladesh, China, India, Nepal, Pakistan and Sri Lanka.

HABITS AND HABITAT Found in a variety of habitats, including arid lands, humid forests and coastal areas. Gleans prey from branches or the ground, including large insects, and small vertebrates, such as lizards, small mammals and birds. Roosts in caves as well as man-made structures.

NOTES This species does not drink blood, unlike true vampire bats from the Americas.

Large-eared Horseshoe Bat ■ *Rhinolophus philippinensis*

HB 5.6–6.1cm, T 2.9–3.7cm, FA 4.7–5.8cm

DESCRIPTION There are about 50 species of horseshoe bat currently recognised from the region, all of which have a distinctive noseleaf that is horseshoe-shaped at the front and more or less pointed at the back. This species can be recognised by its exceptionally large ears and a very long central process in the middle of the nose.

DISTRIBUTION Brunei, Indonesia (Borneo), Malaysia (Sabah, Sarawak), Philippines.

HABITS AND HABITAT Feeds mainly in the understorey of lowland mature forest. Roosts in small groups in caves, often with other species of horseshoe bats. Feeds on insects that it catches in flight.

NOTES The elaborate noseleaf of horseshoe bats helps to direct the echolocation calls which are emitted through the nostrils. Each species in an area usually has a different frequency echolocation call. The species with the largest ears and noseleaf have the lowest frequency calls to detect larger prey.

Intermediate Roundleaf Bat ▪ *Hipposideros larvatus*
HB 6–8cm, T 3–4.5cm, FA 5–6.5cm

DESCRIPTION Dark grey-brown or reddish brown, underparts slightly paler. Ears and noseleaf dark grey or brown. Three lateral leaflets on each side of the noseleaf. The ears are broad and triangular.

DISTRIBUTION Cambodia, Indonesia (Bali, Java, Kalimantan, Sumatra), Laos, Malaysia (Peninsular Malaysia, Sabah, Sarawak), Myanmar, Thailand and Vietnam. Also found in Bangladesh, China and India.

HABITS AND HABITAT Found in a variety of habitats from primary and secondary forests to highly disturbed agricultural land, often associated with limestone caves. Roosts in caves, abandoned mines and rock crevices, sometimes in large numbers.

NOTES This represents a group of similar-looking species, but the distinguishing features and appropriate names for each remain to be clarified.

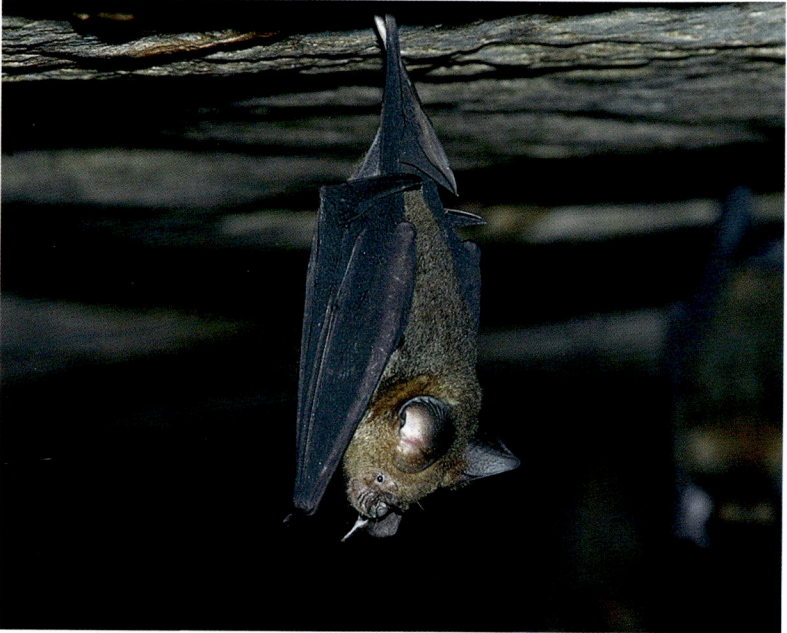

Horsfield's Myotis ■ *Myotis horsfieldii* HB 5.6–6.1cm, T 2.9–3.7cm, FA 4.7–5.8cm

DESCRIPTION Greyish to grey-brown fur, slightly paler below. All Myotis have distinctive oval-shaped ears with a narrow pointed tragus that angles forwards. This species has hind feet somewhat enlarged with wing membrane connected on the side of the foot.

DISTRIBUTION Brunei, Cambodia, Indonesia (Borneo, Java, Sumatra), Malaysia (Peninsular Malaysia, Sabah, Sarawak), Laos, Philippines, Thailand, Vietnam. Also China, India. Further research may show some populations are different species.

HABITS AND HABITAT Roosts during the day in caves or small rock crevices. Feeds over small streams or rivers, catching flying insects or scooping insects off the water with its enlarged feet. Some similar species with even bigger feet are known to sometimes catch small fish.

NOTES *Myotis* is the most widespread naturally occurring genus of mammals, found on all continents except Antarctica. Vespertilionid bats, including *Myotis*, emit echolocation calls through their mouth, so they always fly with their mouth open.

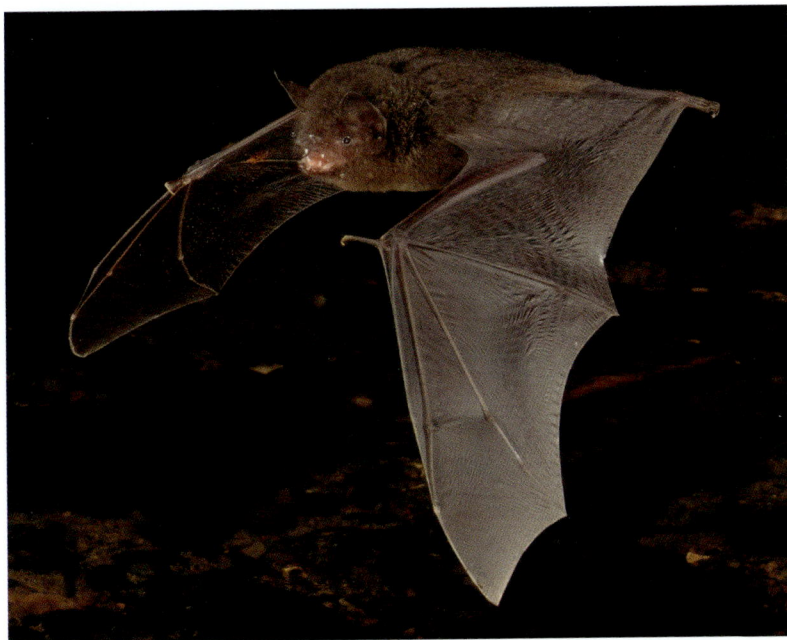

Malayan Greater Bamboo Bat ■ *Tylonycteris malayana*

HB 3.6–3.7cm, T 2.6–2.8cm, FA 2.3–2.7cm

DESCRIPTION Very small with sleek dark brown upperparts, paler below with dark grey wings. Flattened head with short rounded ears. Enlarged rounded pads on the wrists and ankles.

DISTRIBUTION Cambodia, Laos, Malaysia (Peninsular Malaysia), Myanmar, Vietnam. Also found in China and India. The very similar Sunda species, *T. robustula*, is found in Brunei, Malaysia (Sabah, Sarawak) and Indonesia (Borneo, Java, Sumatra).

HABITS AND HABITAT Bamboo bats roost during the day inside the hollow stems of large bamboos. They have flattened skulls enabling them to squeeze through narrow 0.5–0.7cm slits in the bamboo. They are adapted to fly rapidly in open areas, over streams or above the forest canopy, where they catch insects in flight.

NOTES Lesser Bamboo Bats often occur in the same stands of bamboo as this species. Recent genetic research suggests populations of both species found on the Sunda Islands are distinct species from those on the mainland.

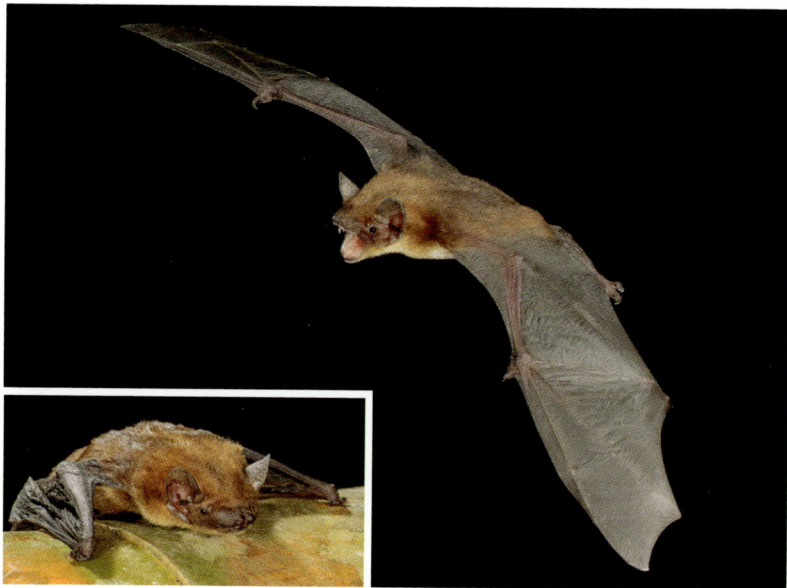

Clear-winged Woolly Bat ∎ *Kerivoula pellucida*

HB 3.5–4.1cm, T 4.0–5.2cm, FA 2.9–3.2cm

DESCRIPTION Long fluffy fur, pale orange-brown above with buffy bases, paler underneath. Wing and tail membranes translucent pale grey with pinkish-orange legs and wing bones. Ears very long with a thin orange tragus.

DISTRIBUTION Brunei, Indonesia (Borneo, Java, Sumatra), Malaysia (Peninsular Malaysia, Sabah, Sarawak), Philippines.

HABITS AND HABITAT Found only in the understorey of tall lowland forest. Roosts singly or in small groups in clumps of dead leaves. Highly manoeuvrable, able to fly slowly or fast and turn quickly. It feeds on insects or other invertebrates that it catches in flight or gleans off leaves or spider webs.

NOTES It has short, very high-frequency echolocation calls, sweeping from about 180kHz to 60kHz, that are well adapted to foraging in cluttered environments.

Lorises are small, stocky nocturnal primates with forward-facing eyes and a vestigial tail. Their hands appear human-like with opposable thumbs. They are related to lemurs, pottos and bushbabies and are considered amongst the more primitive primates.

Philippine Slow Loris ■ *Nycticebus menagensis* HB 26–38cm, T 1–2cm

DESCRIPTION Very small in size. Pale golden to red fur, with usually indistinct facial markings on the round head, though stripes are sometimes darker. Very short ears and large, round eyes.

DISTRIBUTION Brunei, Indonesia (East Kalimantan), Malaysia (Sabah) and the southern Philippines (Bongao, Sangasanga and Tawi Tawi).

HABITS AND HABITAT Found in both primary and secondary lowland forest as well as gardens and plantations. Omnivorous, eating the gum from woody vegetation, as well insects and animal matter. Detailed information on its diet is lacking, but research is ongoing.

NOTES Until recently, it was considered a subspecies of the Sunda Slow Loris (see p. 33).

Javan Slow Loris ■ *Nycticebus javanicus* HB 30–39cm, T 1–2cm

DESCRIPTION Yellowish-brown with a dark dorsal stripe. Head, shoulders and neck are paler. Prominent creamy-white diamond shape between the eyes, formed by a distinct stripe that starts at the top of the head and forks towards the eyes and ears, extending down to the cheeks. Dark fur on ears.

DISTRIBUTION Indonesia (Java).

HABITS AND HABITAT Nocturnal and arboreal, moving slowly between hanging vines and branches like other lorises. Found in both primary and secondary forests. Main diet comprises the gum of trees, as well as insects. Sleeps curled up, hidden in branches.

NOTES Like all species of slow loris in Indonesia, this one is severely threatened by capture for the illegal pet trade. Until recently, also considered a subspecies of the Sunda Slow Loris (see p. 33).

Pygmy Slow Loris ◾ *Xanthonycticebus pygmaeus* HB 21–29cm, T 1–2cm

DESCRIPTION Thick and short, woolly fur that is light brownish-grey to orange and reddish-brown, with silver frosting. Underparts are light grey and ears are relatively conspicuous. Dorsal stripe is faint or absent, and the large eyes are encircled by dark rings. Pygmy Slow Lorises go through seasonal changes, differing greatly in size, colouration and marking between the cold, dry and warm, wet seasons.

DISTRIBUTION Cambodia (east of the Mekong River), Laos and Vietnam. Populations in the northern part of the range are now considered a separate species, *X. intermedius*.

HABITS AND HABITAT Primary evergreen and semi-evergreen forests, limestone forest, and bamboo, secondary and even highly degraded habitats. Forages alone, eating mainly insects and gum, as well as other plant matter and small animal prey such as geckos and birds.

Sunda Slow Loris ■ *Nycticebus coucang* HB 26–30cm, T 1.5–2.5cm

DESCRIPTION Fur colour varies from light grey-brown to reddish-brown, with a dark dorsal stripe that extends from the top of the head to the lower back. On the back of the neck this stripe branches into four lines connecting to the ears and eyes. Dark rings encircle the large eyes.

DISTRIBUTION Indonesia (Sumatra and some offshore islands), Malaysia (Peninsular Malaysia including the islands of Penang, Langkawi and Tioman), Singapore and southern Thailand.

HABITS AND HABITAT Found in primary and secondary lowland forests, as well as gardens and plantations. Diet consists of gum, and also some fruits, insects, birds' eggs and leaves.

Western Tarsier ■ *Cephalopachus bancanus* HB 12–15cm, T 18–22cm

DESCRIPTION Varies in colour, with the subspecies on Borneo (*T. b. bancanus*) being more golden-orange and rusty-brown compared with others that are more ivory-yellow. Tuft of long hair at the end of the long, hairless tail.

DISTRIBUTION Brunei, Indonesia (south Sumatra, Bangka, Belitung, Karimata, Serasen in the South Natuna Islands and Kalimantan) and Malaysia (Sabah and Sarawak).

HABITS AND HABITAT Found in primary and secondary forests, and also forest edges and along coasts. Exclusive diet of animal prey, consuming mainly insects such as beetles, grasshoppers, butterflies, moths and ants. Also eats small vertebrates such as birds, bats and snakes. Infants weigh a mere 25g at birth, but grow rapidly, catching their own insect prey at four weeks of age. Forages low in trees, around 2m off the ground, but if frightened will move higher up.

> **MONKEYS**
> Old World monkeys do not have the prehensile tail found in New World species.
> Monkeys in Southeast Asia are broadly split into colobines (langurs and relatives) and
> macaques.

Thomas's Langur ▪ *Presbytis thomasi* HB 42–61cm, T 50–85cm

DESCRIPTION Langur with striking facial markings. Eyes surrounded by white and grey-blue.
Grey crest flanked by two white stripes, black moustache and flesh-coloured muzzle. Grey
upperparts and white underparts, and a very long, pale tail. Hands and feet are black.

DISTRIBUTION Indonesia (Sumatra – endemic to the northern provinces of Aceh and
North Sumatra).

HABITS AND HABITAT Occurs in primary and secondary rainforests, and neighbouring
rubber plantations. Feeds mainly on young leaves, fruits and flowers, and also on small
animal matter. Lives in family groups, usually of several females and one adult male, or
occasionally two. After reaching maturity, males normally join into small groups of males, or
lead solitary lives. Mainly arboreal, but occasionally descends to the ground to forage.

NOTES There are good chances of observing Thomas's Langurs near the Bohorok Orangutan
Centre in Bukit Lawang and in the Gunung Leuser National Park.

White-thighed Langur ■ *Presbytis siamensis* HB 43–69cm, T 68–84cm

DESCRIPTION Brown to greyish-brown upperparts and tops of the head and arms; black hands, feet and distal half of the tail; and pale grey underside of the body, arms and legs, including a large patch on the outer thighs. Facial skin dark grey to almost black; occasionally, skin around the eyes is paler. Looks very similar to the Common Banded Langur (see p. 37), but that species lacks the pale outer thighs.

DISTRIBUTION Indonesia (Sumatra and Bintan Island in the Riau Archipelago, and possibly Batam and Galang Islands), Malaysia (Peninsular Malaysia) and southernmost Thailand.

HABITS AND HABITAT Found in lowland to hill forests, including disturbed forests and plantations. Diurnal and arboreal.

NOTES This species is easily seen on Fraser's Hill and nearby mountains in Peninsular Malaysia, where it overlaps with Dusky Langurs (see p. 42) and other primate species.

Common Banded Langur ■ *Presbytis femoralis* HB 46–59cm, T 69–77cm

DESCRIPTION Dark-brown to blackish upperparts, and grey underparts with pale patches on the inner thighs. Grey crest with pale grey skin around the eyes.

DISTRIBUTION Indonesia (east and central Sumatra), Malaysia (Peninsular Malaysia in the far south) and Singapore. Populations in northern Peninsular Malaysia, peninsular Myanmar, and peninsular Thailand are now considered a separate species, *P. robinsoni*.

HABITS AND HABITAT Lives in a wide variety of habitats, from mixed mangrove to primary and secondary forests. Like most langurs, its main diet consists of young leaves and fruits.

NOTES This is the only species of langur in Singapore.

Maroon Langur ■ *Presbytis rubicunda* HB 44–58cm, T 67–80cm

DESCRIPTION Reddish-brown to golden-brown fur, and the face has a bluish tinge. Five subspecies exist, with slightly varying fur colouration. *P. r. chrysea*, which occurs in a small area near the Kinabatangan River in Sabah, is paler golden brown than the other subspecies, which are generally more reddish-brown. One of the subspecies occurring in Kalimantan, *P. r. rubicunda*, has blackish extremities to the limbs. The infants are white.

DISTRIBUTION Endemic to Borneo. Occurs throughout most of Borneo, in Indonesia (Kalimantan and Karimata Island) and Malaysia (Sabah and Sarawak). Presence in Brunei uncertain.

HABITS AND HABITAT Largely arboreal, preferring primary and secondary lowland to swamp forests, and also visiting gardens to feed. Consumes a large amount of seeds, and at least some populations are best described as granivorous. Also eats young leaves, fruits and flowers, but the diet varies according to availability of food sources. Lives in groups of up to 13 individuals, and males use loud calls to mark their territories.

Miller's Grizzled Langur ■ *Presbytis canicrus* HB 48–56cm, T 65–84cm

DESCRIPTION Grey upperparts with hairs tipped with white, giving it a grizzled appearance. Underside is light grey to whitish, and undersides of limbs are white. Hands and feet are black, as is the crest. Head is dark grey or blackish. Bare skin on the upper part of the face is dark reddish, turning pinkish at the lower parts of the face.

DISTRIBUTION Endemic to Borneo. Indonesia (eastern Kalimantan).

HABITS AND HABITAT Found in lowland and hill dipterocarp forests, up to 1,600m. While largely arboreal, it does come to the ground, where it frequents mineral springs. Feeds largely on leaves, but also on fruits, seeds and even some animal material such as eggs and nestling birds.

NOTES Once considered a subspecies of Hose's Langur (*P. hosei*), Miller's Grizzled Langur is one of the rarest and least known species of primate in Asia. It has been hunted for the bezoar stones in its gut, which are believed by some to have medicinal properties. Conservation efforts are urgently needed to ensure the survival and recovery of this species.

Javan Grizzled Langur ■ *Presbytis comata* HB 51–54cm, T 65–66cm

DESCRIPTION Grizzled (grey-white mix) coat with a black head and crest, and paler underparts. Slender build, and longer hair frames the broad grey face.

DISTRIBUTION Indonesia (western and central Java).

HABITS AND HABITAT Though in the past it was found across extensive lowlands and mountains, habitat destruction has restricted it to patches of forest in montane habitats. Mainly eats leaves, fruits, flowers and seeds. Lives in troops usually consisting of 7–8 individuals, and occasionally forms groups with the Eastern Ebony Langur, *Trachypithecus auratus.*

NOTES Remaining populations live in some protected areas in western Java, including Ujung Kulon, Halimun and Gede-Pangrango national parks. Found at up to 2,500m asl.

Sundaic Silvered Langur ■ *Trachypithecus cristatus*

HB 41–54cm, T 60–76cm

DESCRIPTION Overall dark grey with pale grey frosting, dark grey face and pointed crest. Infants are bright orange, but patches of grey appear until they develop the full adult colouration. Long limbs and very long tail.

DISTRIBUTION Brunei, Indonesia (Kalimantan, Sumatra, Bangka, Belitung and the Riau and Lingga archipelagos off eastern Sumatra, as well as the Natuna Islands), and Malaysia (Sabah and Sarawak).

HABITS AND HABITAT Found in coastal, riverine and mangrove swamp forests. Diet comprises mainly young leaves, shoots, flowers, seeds and fruits, especially those of mangrove species. Lives in groups of 10–50 individuals.

Dusky Langur ■ *Trachypithecus obscurus* HB 50–70cm, T 70–80cm

DESCRIPTION Distinctive face with incomplete white rings around the eyes against dark grey facial skin, giving it the appearance of spectacles, hence its other common name, the Spectacled Leaf Monkey. Greyish-brown to dark grey upperparts with paler grey outer hind legs, tail and crest. Bare pink patches on the lips. Infants are light orange.

DISTRIBUTION Thai-Malay Peninsula: Malaysia (Peninsular Malaysia including associated islands), Myanmar (south) and Thailand (south-western).

HABITS AND HABITAT Inhabits primary and secondary forests, from lowlands to mountains, up to 1,800m. Prefers old-growth forests, but can be found in a wide range of disturbed habitats. Diet of mainly young leaves, shoots, flowers, seeds and fruits, especially those of mangrove species. Lives in groups of 10–50 individuals.

Hatinh Langur ■ *Trachypithecus hatinhensis* HB 50–66cm, T 81–87cm

DESCRIPTION Glossy black fur overall, except for a white moustache extending from the sides of the mouth over the ears to the nape. Distinct black crest. Juveniles have a white band on their foreheads. Fairly similar to Francois' Langur (*Trachypithecus francoisi*) and the Lao Langur (*T. laotum*), but the distinguishing characteristics are that the former's moustache does not continue behind its ears and the latter has a white forehead even in adults.

DISTRIBUTION Laos (east-central) and Vietnam (north-central).

HABITS AND HABITAT Spends considerable time in trees and on the ground. Mainly eats leaves. Lives in forests near limestone karsts and outcrops in rocky mountainous areas, though its previous range may have included a wider habitat variety.

Capped Langur ▪ *Trachypithecus pileatus* HB 50–70cm, T 80–100cm

DESCRIPTION Distinct black cap and dark face; facial hair and underparts pale yellow to orange. Hairs on the crown are short, sticking straight up. Grey to pale brown body, and tail with black tip. Easy to distinguish from other langurs as none of the others have such a dark cap contrasting with a yellow-orange face. Juveniles up to the age of five months are creamy-white with a pink face.

DISTRIBUTION Myanmar (north-west). Also in Bangladesh (east), Bhutan and India (north-east).

HABITS AND HABITAT Found in evergreen, semi-evergreen, moist deciduous, bamboo and open woodlands. Diurnal and largely arboreal. Eats mainly leaves and also fruits, seeds and flowers. Lives in groups of multiple females and one male.

Red-shanked Douc ▪ *Pygathrix nemaeus* HB 61–76cm, T 56–76cm

DESCRIPTION A strikingly colourful, large primate. Dark reddish-chestnut lower legs, and black hands, feet, shoulders, insides of upper arms, upper legs and rump; white lower arms and speckled grey back, belly and tops of the upper arms. Yellow-brown face and a long white tail.

DISTRIBUTION Cambodia (small area in the north-east), Laos (east-central and south-east), Vietnam (north and central – very fragmented).

HABITS AND HABITAT Found in tall evergreen and semi-evergreen primary forests, and lowlands up to 2,000m, including limestone outcrops. Mainly arboreal, and feeds on leaves and buds, as well as some fruits, flowers and seeds. When relaxed, it moves noisily in the trees, disappearing quietly only when disturbed.

Black-shanked Douc ▪ *Pygathrix nigripes* HB 61–76cm, T 56–76cm

DESCRIPTION Dark-speckled grey crown and upperparts, paler grey underparts, white chin, throat and tail, and black limbs with paler frosting on the arms. Blue-grey facial skin with distinct yellow-orange eye-rings.

DISTRIBUTION Cambodia (east) and Vietnam (south-west), and possibly Laos.

HABITS AND HABITAT Lives in evergreen, semi-evergreen and mixed deciduous forests. Feeds on leaves, seeds, fruits, flowers and buds.

Proboscis Monkey ■ *Nasalis larvatus* HB 55–65cm, T 62–74cm

DESCRIPTION Best known for its oversized nose, the Proboscis Monkey is a very large primate, with males reaching up to 20kg in weight. Adult males have large stomachs. Reddish-brown fur and greyish limbs. Females are significantly smaller than males, and like the juveniles, they have small, upturned noses.

DISTRIBUTION Endemic to Borneo. Brunei, Indonesia (Kalimantan) and Malaysia (Sabah and Sarawak).

HABITS AND HABITAT Inhabits riparian-riverine forests and coastal lowland forests, including mangrove, peat-swamp and freshwater swamp forests. Mainly eats young leaves and unripe fruits. It has partly webbed back feet, which aid it in balancing on mangrove mud and swimming.

Southern Pig-tailed Macaque ■ *Macaca nemestrina*

HB 47–59cm, T 14–23cm

DESCRIPTION Stocky, heavy-set macaque with a short, curly tail and an olive-brown coat, a dark brown crown and whitish underparts.

DISTRIBUTION Brunei, Indonesia (Kalimantan and Sumatra), Malaysia and Thailand (southern peninsula).

HABITS AND HABITAT Often found in hilly areas, foraging largely on the ground. Diurnal and widespread. Eats fruits and small animals. Usually lives in large groups of 15–40 individuals, but males are sometimes solitary.

Long-tailed Macaque ■ *Macaca fascicularis* HB 45–55cm, T 44–55cm

DESCRIPTION Grey-brown to reddish-brown fur, with slightly paler undersides and a brownish-grey face. Lean build, with males being significantly larger than females, and a long tail.

DISTRIBUTION Brunei, Cambodia, Indonesia (Sumatra, Java, Bali and most but not all offshore islands in the Greater Sundas), Laos (southern), Malaysia, Myanmar (south), the Philippines, Singapore, Thailand (west-central, east and south, including offshore islands), Timor-Leste and Vietnam (south-east). Also in Bangladesh (south-west) and India (Andaman and Nicobar Islands).

HABITS AND HABITAT Especially common near coastal areas and forest edges, frequently near people. An omnivore, it eats a wide range of animal matter and vegetation. It is diurnal, sleeping in the branches of trees during the night. Gregarious, often congregating in groups of 20–30 individuals, though sometimes groups number more than 50.

NOTES This is one of a few primate species that use tools: populations in southern Thailand and southern Myanmar use stones and shells to crack open marine molluscs and nuts. The only macaque in the Philippines.

Siamang ■ *Symphalangus syndactylus* HB 75–90cm, T absent

DESCRIPTION The largest of the gibbons, this species has a stocky build relative to other gibbon species. It has shaggy black hair, and a greyish lower face. Both males and females have a throat pouch that visibly inflates when calling. Males also have a long scrotal tuft that resembles a short tail.

DISTRIBUTION Indonesia (Sumatra), Malaysia (Peninsular Malaysia) and Thailand (southernmost, in the Thai-Malay Peninsula).

HABITS AND HABITAT Found in primary and secondary forests, from lowlands up to 1,500m, in small family groups of a pair and offspring. Arboreal but moves less gregariously than other gibbons, instead brachiating more gracefully through the trees. Siamangs have a loud, booming call, followed by whooping, and males end their calls with a loud yell.

NOTES Lives sympatrically with other gibbons, that is with the White-handed or Agile Gibbons (see pp. 51 and 52) in Sumatra and Peninsular Malaysia.

White-handed Gibbon ■ *Hylobates lar* HB 45–60cm, T absent

DESCRIPTION Blond or dark-brown forms can occur in the same family, unrelated to sex. It has long limbs, with white feet and hands. There is a pale ring encircling its face.

DISTRIBUTION Indonesia (north Sumatra), northern Laos (north-west of Mekong), Malaysia (Peninsular Malaysia), Myanmar (south and east) and Thailand (south-west and north-west). Also possibly in China (south – possibly extirpated).

HABITS AND HABITAT Often seen dangling from branches or huddled in forks of trees. Loud and distinct, high-pitched, whooping call. Mainly eats fruits, young shoots, leaves and insects. An important seed disperser as it swallows nearly all the seeds in its food.

Agile Gibbon ▪

Hylobates agilis
HB 45–65cm, T absent

DESCRIPTION Occurs in two colour forms, buff-blond and very dark brown, the latter more common in mainland Southeast Asia. It has a pale brow and cheeks, with a dark face.

DISTRIBUTION Indonesia (Sumatra – south-east of Lake Toba and the Singkil River), Malaysia (Peninsular Malaysia – from the Mudah and Thepha Rivers in the north to the Perak and Kelanton Rivers in the south) and Thailand (southernmost Thailand, near the Malaysian border).

HABITS AND HABITAT Lives in small family groups in tall dipterocarp forests; rarely descends to the ground.

NOTES Severely threatened by habitat loss due to logging and conversion to oil palm, and by illegal wildlife trade.

Müller's Bornean Gibbon ▪ *Hylobates muelleri*
HB 42–47cm, T absent

DESCRIPTION Fur colour varies from grey to brown. Dark-coloured cap and dark chest and underparts. Pale cream brows, which sometimes extend to appear like an incomplete facial ring.

DISTRIBUTION Indonesia (south-east Kalimantan). Closely related gibbons found elsewhere in Borneo are now considered distinct species.

HABITS AND HABITAT Found in primary and secondary forests, and in selectively logged forest. Favours high-sugar, fleshy fruits, but also eats young leaves and small insects.

◾ GIBBONS ◾

Western Hoolock Gibbon ◾ *Hoolock hoolock* HB 45–65cm, T absent

DESCRIPTION Generally shaggy haired. Adult males and juveniles are mostly black, with thick, joined white eyebrows, and a white tuft on the chin or under the eyes. The preputial tuft is black or faintly grizzled. Females are coppery-buff-brown, with a white facial ring and dark brown cheeks.

DISTRIBUTION Myanmar (north-west). Also in Bangladesh and India (north-east).

HABITS AND HABITAT HABITS Found in primary and semi-evergreen forests. Fruits, leaves and shoots are the main diet. Females lead the movement of the group, which sometimes travels on the ground to reach fruiting trees, especially in degraded habitat

NOTES Formerly conspecific with the Eastern Hoolock Gibbon (*H. leuconedys*).

ORANGUTANS
Orangutans are Asia's only great apes. The Sumatran Orangutan was formerly considered a subspecies of the Bornean Orangutan, but recent research has concluded that Borneo holds one species and Sumatra another.

Bornean Orangutan ◼ *Pongo pygmaeus* HB 780–970cm, T absent

DESCRIPTION Large bodied and generally heavier set than the Sumatran Orangutan (see p. 55), and has darker reddish-brown hair that varies from deep orange to maroon-brown. It also has a broader face with less facial hair than the Sumatran Orangutan.

DISTRIBUTION Endemic to Borneo. Indonesia (south-west, central, east and north-west Kalimantan) and Malaysia (Sabah, north and south Sarawak). Not known to be resident in Brunei.

HABITS AND HABITAT Lives mainly in lowland rainforests, and swamp and mountain forests. Peat swamps and forests that are prone to flooding produce larger and more regular fruit crops, and therefore hold the highest densities of orangutans.

NOTES There are three subspecies on Borneo. The north-west subspecies *P. p. pygmaeus*, with approximately 1,500 remaining individuals, is the most threatened mainly because of severe loss and fragmentation of its habitat, and hunting.

Sumatran Orangutan ▪ *Pongo abelii* HB 780–970cm, T absent

DESCRIPTION Large-bodied ape with a reddish-brown shaggy coat, long limbs and no tail. It has bare, dark facial skin, though juveniles have pale pinkish facial skin. Males are much larger than females. Though generally very similar in appearance to the Bornean Orangutan (see p. 54), Sumatran Orangutans are thinner, and have a lighter red coat, longer hair and longer faces.

DISTRIBUTION Indonesia (north-west Sumatra).

HABITS AND HABITAT Almost completely arboreal – only adult males descend to the ground and do so very rarely. In the wild, it is estimated that males can live to up to 58 years of age, and females to 53.

NOTES Orangutans in Suaq Belimbing, Sumatra, are the only ones known to regularly make and use tools from sticks to extract termites and honey from hard-to-reach spots inside tree trunks. Though this kind of tool use has yet to be observed in Borneo, all orangutans use some tools to an extent, such as large leaves as umbrellas.

Golden Jackal ▪ *Canis aureus* HB 60–80cm, T 20–25cm

DESCRIPTION Usually golden brown to tan with black-tipped hairs on the shoulders and back. The tail is bushy with a dark tip and hangs straight down, lacking the tip-curl usually seen on even the most jackal-like domestic dog. Golden Jackals have pointed ears, and blunt nails, having feet much like a domestic dog.

DISTRIBUTION Myanmar, Thailand, Cambodia, Laos and Vietnam. Also widespread through north and northeast Africa, down the Arabian Peninsula, in parts of Europe, through Central Asia and the entire Indian sub-continent, including Sri Lanka.

HABITS AND HABITAT Found in a wide variety of habitats throughout its range. In Southeast Asia, it occupies lowland, open, decidous forest and grasslands. It can cope with some human disturbance, although in many parts of its range it is hunted as a pest or caught in indiscriminate snares and traps.

NOTES Most often seen along forest edges, in clearings or along trails. In areas where prey is abundant and threats are minimal, it occurs at higher densities and may be easier to see.

Dhole ▪ *Cuon alpinus* HB 80–105cm, T 30–45cm

DESCRIPTION Reddish brown with a paler underside and a bushy tail that becomes increasingly dark to black towards the tip. The ears are rounded with white hairs inside. This is the largest of Southeast Asia's wild dog species.

DISTRIBUTION Cambodia, Indonesia (Java, Sumatra), Laos, Malaysia (Peninsular Malaysia), Myanmar, Thailand and Vietnam. Also found in India, north to Russia (southern Siberia). Current distribution is highly fragmented, and this species is now absent from many parts of its historical range.

HABITS AND HABITAT Found living and hunting in packs, it uses a variety of habitats, including primary and secondary forests, grassland–scrub–forest mosaics and montane forests. It hunts a wide variety of prey, including large mammals such as deer.

NOTES Very little is known about its status, distribution and ecological needs in Southeast Asia, and more research is needed. Habitat loss, loss of prey, persecution and accidental snaring/trapping are some of the main threats to the continued survival of this beautiful dog.

Sun Bear ◼ *Helarctos malayanus* HB 110–140cm, T 3–7cm

DESCRIPTION The Sun Bear is the smallest of the world's bears. It has dark brown to black short fur, with a lighter brown muzzle and a large white to yellowish marking, often V-shaped, on the chest. This marking varies from one individual to the next. Ears are small and rounded. The claws are very long, powerful and non-retractable. Females are generally 10–20% smaller than males.

DISTRIBUTION Brunei, Cambodia, Indonesia (Kalimantan, Sumatra), Laos, Malaysia (Peninsular Malaysia, Sabah, Sarawak), Myanmar, Thailand and Vietnam. Also found in Bangladesh, China and India. Extirpated in Singapore.

HABITS AND HABITAT In Indonesia (Sumatra) and Malaysia, it is found in dense tropical rainforest, including lowland forests, swampy areas, hills, limestone karst hills, and lower montane forest. While it can use selectively logged forests and other disturbed areas, there is no evidence that it can survive in deforested or agricultural land. It climbs extremely well and frequently builds nests in trees to sleep in. It does not hibernate.

NOTES The Sun Bear has been extirpated from many parts of its range, largely due to hunting and habitat loss, and its distribution has become increasingly patchy. The two greatest threats are habitat loss and hunting for commercial trade in its body parts, which are used in traditional Asian medicines. Further research is needed to assess the current status and conservation needs of this species. Reducing the trade in bear parts is of paramount importance.

Asian Black Bear ▪ *Ursus thibetanus* HB 120–150cm, T 6–10cm

DESCRIPTION Usually uniformly black, though sometimes brown (a golden-brown colour phase has been observed in Cambodia), with a lighter brown muzzle and a distinctive creamy-white to yellowish marking across its chest, extending from shoulder to shoulder. Its coat is somewhat shaggy in appearance, compared to the shorter fur of the Sun Bear, especially on its neck. It has relatively large rounded ears, which are much larger than the Sun Bear's. Males are considerably larger than females.

DISTRIBUTION Cambodia, Laos, Myanmar, Thailand and Vietnam. Also found in Afghanistan, Bangladesh, Bhutan, China, India, Iran, Japan, North and South Korea, Nepal, Pakistan, Russia and Taiwan. It occupies all countries in mainland Southeast Asia except Malaysia. There is wide overlap in Southeast Asia with the Sun Bear (see p. 58).

HABITS AND HABITAT Found in a variety of habitats, including both broadleaf and coniferous forests. Also uses secondary forests, taking advantage of fruit, berries or young bamboo shoots and other new growth. Occasionally visits agricultural areas and fruit orchards.

NOTES Threatened by habitat loss and hunting for their body parts, especially paws and gall bladders. Trade in live individuals for pets and for bear bile extraction is also a threat.

Eastern Red Panda ■ *Ailurus styani* HB 51–63cm, T 28–48cm

DESCRIPTION The Red Panda has long soft fur. Its upperparts are reddish, darkest in the middle of the back, with paler underparts. The fur on the head is paler, with white rims to the ears and dark marks under the eyes. The long bushy tail is banded with inconspicuous rings.

DISTRIBUTION Myanmar. Also found in China. Recently split from *Ailurus fulgens* that occurs farther west in Bhutan, China, India and Nepal.

HABITS AND HABITAT Found mainly in montane temperate forests at 1,800–4,000m with bamboo thicket understoreys. It is primarily nocturnal, sleeping on tree branches during the day. It is an efficient climber, feeding in the trees as well as on the ground, on a variety of vegetation including bamboo shoots, grasses, fruit, roots and acorns as well as insects, eggs and small invertebrates. Bamboo leaves are an especially important source of food during the winter months.

Martens
Martens are medium-sized carnivores that are closely related to weasels, otters, badgers, ferrets, minks and stoats.

Yellow-throated Marten ■ *Martes flavigula* HB 45–65cm, T 37–45cm

DESCRIPTION A long, slender body and a long tail. There are variations in coat pattern depending on geographic location. Its back colour varies from medium brown to pale yellowish brown, with darker brown to black lower back, hind legs, lower half of forelegs and tail. It has a black stripe on the side of its neck behind its ear, while the top of its head varies from black to dark brown, extending from the neck stripe. It has a pale yellowish chin, throat and chest with a darker lower belly.

DISTRIBUTION Brunei, Cambodia, Indonesia (Java, Kalimantan, Sumatra), Laos, Malaysia (Peninsular Malaysia, Sabah, Sarawak), Myanmar, Thailand and Vietnam. Also found in Bangladesh, Bhutan, China, India, Nepal, Pakistan, Russia, Taiwan, and North and South Korea.

HABITS AND HABITAT Mainly diurnal but also occasionally hunts at night. Found in a variety of habitats, both natural and disturbed. Its diet comprises a variety of animal matter including birds, snakes, lizards, insects, as well as fruit, nectar and honey. Often solitary or in pairs, but also seen in small family groups.

NOTES Its graceful bounding movements are distinctive; it moves with agility on the ground and in the trees.

> **BADGERS**
> Badgers are generally stocky-bodied with short legs made for digging. Ferret badgers are
> the smallest members of the badger family.

Large-toothed Ferret Badger ■ *Melogale personata*

HB 33–39cm, T 14–21cm

DESCRIPTION The Large-toothed Ferret Badger has brownish to greyish upperparts, paler
underparts, occasionally with an orange tinge. Sides of the body frosted with white. White
dorsal stripe extending from the back of the neck to the middle of the back, sometimes to the
rump or tail. Tail is bushy and pale in colour, with the distal half being white. The head has
a distinctive dark-brown to black and white pattern, and it has massive teeth. Very similar to
other ferret-badgers, and difficult to distinguish in the field.

DISTRIBUTION Cambodia, Laos, Myanmar, Thailand and Vietnam. Also found in China
and India.

HABITS AND HABITAT Nocturnal and terrestrial, although it occasionally climbs trees. Found
in forests, grasslands, and sometimes cultivated areas. Omnivorous, feeding on a variety of
invertebrates, frogs, and sometimes carcasses of small birds and mammals, eggs and fruit.

Sunda Stink-badger ■ *Mydaus javanensis* HB 37–52cm, T 3–5cm

DESCRIPTION It has a black coat with a white dorsal stripe. The extent of the stripe varies, but it generally extends from the top of its head to its tail. It has a long muzzle and a very short tail.

DISTRIBUTION Indonesia (Java, Kalimantan, Sumatra) and Malaysia (Sabah, Sarawak). Presence in Brunei uncertain.

HABITS AND HABITAT Found in secondary forests and areas adjacent to forests. Its diet includes eggs, carrion, insects, earthworms, larvae as well as plants. It is terrestrial and nocturnal, and sleeps in burrows during the day. It uses its long muzzle and long claws to dig into soft soil in search of food.

NOTES The Sunda Stink-badger emits a strong odour from secretions of its anal gland, hence its name. Not a badger at all, this species and its close relative, the Palawan Stink-badger *Mydaus marchei* are the only two species of skunks (Mephitidae) in the Old World.

> **OTTERS**
> Otters are semi-aquatic carnivorous mammals.

Smooth Otter ■ *Lutra perspicillata* HB 65–75cm, T 40–45cm

DESCRIPTION The largest of Asia's four otter species, the Smooth Otter is long and sleek. Upperparts are brown, underparts slightly lighter. The throat and sides of the neck, extending from the chin to the chest, are creamy. The tail is flattened – much more so than in other Asian otter species. The head appears blunt compared with Asia's other larger otter species, and the nose is hairless. Large webbed feet.

DISTRIBUTION Brunei, Cambodia, Indonesia (Java, Kalimantan, Sumatra), Laos, Malaysia (Peninsular Malaysia, Sabah, Sarawak), Myanmar, Singapore, Thailand and Vietnam. Also found in Bangladesh, Bhutan, China, India, Iraq, Nepal and Pakistan.

HABITS AND HABITAT Diurnal, found in primary and disturbed habitats, in coastal areas such as estuaries and mangroves, and in lakes, reservoirs, ponds and large rivers. Sometimes solitary, although more often in groups of up to 15 or more, with 4–6 being the norm. Hunts cooperatively, taking a wide variety of prey including fish, crustaceans, shellfish, amphibians, reptiles and even some small mammals.

NOTES One of the more easily observed species in some parts of its range, such as Malaysia and Singapore. Various birds often follow groups of Smooth Otters as they hunt, taking advantage of the disturbed fish and the chance of an easy meal.

Oriental Small-clawed Otter ■ *Lutra cinerea* HB 36–55cm, T 22–35cm

DESCRIPTION The smallest of the region's four otter species. Upperparts dark brown, underparts paler brown. The sides of neck and throat, the chin and cheeks are a buffy colour. The muzzle is blunt. As its name implies, it has short claws, not extending beyond the ends of the digits.

DISTRIBUTION Brunei, Cambodia, Indonesia (Java, Kalimantan, Sumatra), Laos, Malaysia (Peninsular Malaysia, Sabah, Sarawak), Myanmar, Philippines (Palawan), Singapore, Thailand and Vietnam. Also found in Bangladesh, Bhutan, China, India, Nepal and Taiwan.

HABITS AND HABITAT Largely diurnal. Occurs in a wide variety of habitats with permanent water and forest cover, including sea coasts, mangroves, rivers, streams, ponds and lakes, as well as in agricultural paddy areas. Mostly in groups of 4–15 animals, though there is little information on group composition or behaviour.

NOTES As with the other Asian otters, populations have declined throughout much of its range due to habitat loss and hunting for trade in fur, meat and body parts used in traditional medicines.

Hairy-nosed Otter ■ *Lutra sumatrana* HB 55–72cm, T 37–48cm

DESCRIPTION Large, brown with creamy chin, lips and upper parts of the throat. It has prominent claws and fully webbed digits. The head is flatter and less blunt than that of the Smooth Otter (see p. 64). The unique hair-covered nose, which gives it its common name, is one of the most important features to distinguish this species from the very similar Eurasian Otter (see p. 67).

DISTRIBUTION Brunei, Cambodia, Indonesia (Sumatra, Kalimantan), Malaysia (Peninsular Malaysia, Sabah, Sarawak), Thailand and Vietnam. Presence in Laos and Myanmar uncertain.

HABITS AND HABITAT Very little is known of this rare nocturnal species. It appears to be solitary in nature, but it may occur in groups of up to six. Found in coastal areas, swamps, large rivers and associated tributaries and lakes.

NOTES The Hairy-nosed Otter is severely threatened by hunting for skins, meat and medicine. More research is essential to better understand and protect this species.

Eurasian Otter ■ *Lutra lutra* HB 50–80cm, T 37–50cm

DESCRIPTION Brown upperparts with paler brown underparts. The chin and throat are lighter brown. Very similar in appearance to the Hairy-nosed Otter (see p. 66), but the tip of the nose is hairless. The coat appears grizzled and rough. The head is flatter and less blunt than in the Smooth Otter (see p. 64).

DISTRIBUTION Cambodia, Indonesia (Sumatra), Laos, Myanmar, Thailand and Vietnam. Also found from Europe, through North Africa, North Asia, South Asia and East Asia.

HABITS AND HABITAT Largely nocturnal, although sometimes active during the day, depending on the vulnerability of prey in daylight hours. Occurs in a wide variety of habitats, including highland and lowland lakes, rivers, streams, swamps and coastal areas, from sea-level to high in the mountains. Usually near banks with substantial covering vegetation.

While fish make up the bulk of the diet, it also eats crustaceans, aquatic insects, reptiles, amphibians, birds and small mammals. Largely solitary, although sometimes in small groups made up of mother and her offspring.

NOTES In some parts of its range the distribution of the Eurasian Otter overlaps with that of Smooth Otters and Oriental Small-clawed Otters (see p. 65).

Large Indian Civet ▪ *Viverra zibetha* HB 75–85cm, T 38–46cm

DESCRIPTION The Large Indian Civet is grey-brown with mottled dark spots that turn into wavy lines closer to the rump. Black erectile hairs run from the back of the neck down the back, ending at the base of the tail. The thick tail is ringed with five or six white bands, alternating with black bands. The neck is boldly marked with black and white. A sheath of skin covers the claw on the third and fourth toe.

DISTRIBUTION Cambodia, Laos, Malaysia (Peninsular Malaysia), Myanmar, Thailand and Vietnam. Also found in Bhutan, China, India and Nepal. Possibly extirpated in Singapore.

HABITS AND HABITAT Found in primary and secondary forests, and occasionally in plantations or otherwise altered and degraded habitats. It has been recorded up to 1,600m, rarely below 400m, where a related species, Large-spotted Civet *Viverra megaspila*, tends to predominate. While largely terrestrial, it can climb. Feeds on a wide variety of prey, including fish, birds, lizards, frogs, insects, arthropods and crabs, as well as domestic poultry and even garbage.

NOTES While this species appears to be common across much of its range, hunting, trapping and snaring are apparently taking a toll, and as a result it is becoming increasingly rare, especially where the threat of trapping goes hand-in-hand with extensive habitat loss and fragmentation. In some areas, it is now completely absent.

Malay Civet ▪ *Viverra tangalunga* HB 61–67cm, T 28–36cm

DESCRIPTION Upperparts are greyish with numerous black spots and a black dorsal stripe that runs to the tip of the tail. Underparts are whitish and there are bold black markings on the otherwise white throat. The legs are dark to black. There are about 15 black bands on the tail. Longish legs and a pointed muzzle.

DISTRIBUTION Indonesia (Kalimantan, Sumatra), Malaysia (Peninsular Malaysia, Sabah, Sarawak), Philippines and Singapore. Presumably in Brunei.

HABITS AND HABITAT Largely nocturnal, but may be crepuscular. Terrestrial, although they can climb. Found in a variety of habitats including primary and secondary forests, cultivated land, plantations and near human settlements in the proximity of forest, over a wide altitudinal range, from sea level to at least 1,200m. It is largely solitary and has an omnivorous diet. Rests during the day at ground level hidden in logs, dense brush piles or thick vegetation.

NOTES May be seen at forest edges, often near agricultural areas, orchards and gardens, in the late evening, as it forages on the ground.

Masked Palm Civet ■ *Paguma larvata* HB 51–76cm, T 51–64cm

DESCRIPTION Colour is variable throughout the range of this species, ranging from dark brown to reddish to light brown. The tail is often dark with a white tip, but not always. It has a dark mask, ears, muzzle and legs and has white cheeks. In some populations, especially in the northern parts of Southeast Asia, there is white on the top of the head running from the nose to the nape, although this is less conspicuous further south, and absent altogether in the extreme south of its range.

DISTRIBUTION Brunei, Cambodia, Indonesia (Kalimantan, Sumatra), Laos, Malaysia (Peninsular Malaysia, Sabah, Sarawak), Myanmar, Thailand and Vietnam. Also found in Bhutan, China, India (Andaman Islands) and Nepal.

HABITS AND HABITAT Nocturnal with occasional diurnal activity, and partially arboreal. Occurs in primary and secondary forests, including peat-swamp forests and disturbed areas, up to 2,500m. The omnivorous diet includes small mammals, insects and fruits. Females have up to four young per litter, with two breeding seasons per year.

NOTES This species can sometimes be observed foraging on the ground, often on trails near human settlements, in areas where hunting is minimal.

Common Palm Civet ▪ *Paradoxurus hermaphroditus*

HB 42–50cm, T 33–42cm

DESCRIPTION It has a long body and short dark legs, with a dark long tail. The upperparts vary widely in colour and pattern, but often are greyish brown with three dark broken lines running down the back, and irregular dark spots along the sides, sometimes forming lines as well. The underparts are lighter grey. A broad mask-like band covers the face, including the base of the pointed muzzle and ears. The forehead is lighter grey to whitish, as are the cheeks and the fore-part of the muzzle.

DISTRIBUTION Recent research suggests this is a complex of species with *P. philippinensis* in Philippines, Brunei, Indonesia (Kalimantan) and Malaysia (Sabah, Sarawak); *P. musangus* in Indonesia (Sumatra, Java), Peninsular Malaysia, Singapore, Thailand, Cambodia, Laos and Vietnam; and *P. hermaphroditus* in Cambodia, Laos, Myanmar, Thailand and Vietnam. Also in Bangladesh, Bhutan, China, India, Nepal and Sri Lanka, but species limits poorly known.

HABITS AND HABITAT Largely arboreal, crepuscular and nocturnal, it is found in a wide range of habitats including evergreen and deciduous forest (primary and secondary), as well as plantations, suburbs and even urban green-space, up to 2,400m. Adapted for forest living, yet often found in areas near humans; sleeping in barns, drains, or roofs during the day, and emerging at night to catch rodents or forage for mango, coffee, pineapples, melons and bananas. It also eats insects and molluscs.

NOTES In some parts of its range this species is hunted for meat, captured for the 'civet coffee' trade and the pet trade, and also persecuted as a pest.

Banded Civet ■ *Hemigalus derbyanus* HB 45–56cm, T 25–36cm

DESCRIPTION This beautiful civet has a light golden-brown to pale brown body with distinct broad dark brown to black bands across its back and at the base of its tail. The tail is darker than the body, especially further from the base. Longitudinal stripes on the neck and to the face. The underside is lighter and without barring. Pointed face with long whiskers.

DISTRIBUTION Indonesia (Kalimantan, Sumatra), Malaysia (Peninsular Malaysia, Sabah, Sarawak), southern Myanmar and southern Thailand. Presumably in Brunei as well. Appears to be more common on Borneo than elsewhere in Southeast Asia.

HABITS AND HABITAT This nocturnal ground-dwelling species has been recorded from primary lowland rainforest, but also in disturbed habitat, peat-swamp forest and acacia plantations. In Borneo, it has been found at elevations up to 1,200m. Significantly less abundant in secondary or disturbed forests, and there is no clear evidence that the Banded Civet can persist in plantations.

NOTES Further study of the ecology of this species is required. Being mainly terrestrial, these animals are at risk from snares and other traps.

Owston's Civet ■ *Chrotogale owstoni* HB 51–63cm, T 38–48cm

DESCRIPTION Pale brown to light grey in colour with broad black bands across the body and the base of the tail. The underside is orangeish in males and paler, yellowish, in females. The tail is darker than the body, especially further from the base. Longitudinal stripes on the neck and face, and dark brown to black spots on the sides and down the legs. The underside is lighter and lacks barring. Elongated face with long whiskers and pointed ears.

DISTRIBUTION Laos and Vietnam. Presence in Cambodia uncertain. Also found in southern China.

HABITS AND HABITAT While little is known of this species' habitat use and general ecology, it has been recorded in lowland and montane evergreen forests, broadleaf forests over limestone, and bamboo forest. Often found in mountainous highlands, and has also been recorded from heavily degraded forest and at forest edges. The range restriction of this species in eastern Laos may be associated with the extent of wet evergreen forest in the eastern Annamites.

NOTES Little is known about the range and ecology of Owston's Civets, and further studies are required. The species is severely threatened by hunting, and is particularly vulnerable to ground-level snares and traps.

Otter Civet ■ *Cynogale bennettii* HB 57–68cm, T 12–21cm

DESCRIPTION This civet has a dark brown coat, with faint grey grizzling and pale underparts. It has prominent white lips, very long whiskers, small ears and a faint pale spot above each eye. It has partially webbed feet.

DISTRIBUTION Brunei, Indonesia (Kalimantan, Sumatra), Malaysia (Peninsular Malaysia) and southern Thailand.

HABITS AND HABITAT This is a poorly known species, but it is thought to be largely confined to peat-swamp forests, though there are recent records from lowland dry forest. It seems to prefer lowland primary forest, and is also found in secondary forest, bamboo, and logged forest, but details on its long-term habitat use are lacking. It is semi-aquatic, assumed to hunt in and near water for fish, crabs, molluscs as well as birds and small mammals. Thought to be primarily nocturnal.

NOTES Very little is known about this species, from its ecology to its population trends, pointing to an urgent need for further study.

Binturong ■ *Arctictis binturong* HB 65–95cm, T 50–80cm

DESCRIPTION Shaggy black hair with white frosting covers this long and low-slung civet. The head is slightly paler and has round, tufted ears. It has an extremely large, thick, prehensile tail. Sometimes unfortunately referred to as a 'bearcat', although is not closely related to either bears or cats.

DISTRIBUTION Brunei, Cambodia, Indonesia (Java, Kalimantan, Sumatra), Laos, Malaysia (Peninsular Malaysia, Sabah, Sarawak), Myanmar, Philippines (Palawan), Thailand and Vietnam. Also found in Bangladesh, Bhutan, China, India and Nepal. Extirpated in Singapore. Historically abundant but now uncommon or rare over much of the range.

HABITS AND HABITAT Primarily arboreal. Due to its heaviness and sluggish nature it cannot leap canopy gaps, so it often has to descend to the ground to move through the forest. Crepuscular and nocturnal, although sometimes active during the day. Confined to tall forest, where it feeds on fruits and small animals such as insects, birds, rodents and fish. In the Philippines, the species is found in primary and secondary lowland forest, including grassland–forest mosaic.

NOTES Nine subspecies have been described. Of those currently recognised, the Palawan Island population (*A. b. whitei*) is sometimes considered a distinct species. Threatened by habitat loss, but also hunted for pets and consumption.

Small-toothed Palm Civet ■ *Arctogalidia trivirgata*

HB 44–53cm, T 48–66cm

DESCRIPTION Long in appearance, with a tail longer than the body, and short legs. It is usually dark greyish brown, though sometimes slightly lighter, with slightly lighter and reddish underparts. Three black stripes run down the back. These stripes can be difficult to see, and care must be taken not to confuse this species with Common Palm Civets. The face, ears, tail and feet are also black. There is a light stripe running from the forehead to the tip of the muzzle.

DISTRIBUTION Brunei, Cambodia, Indonesia (Java, Kalimantan, Sumatra), Laos, Malaysia (Peninsular Malaysia, Sabah, Sarawak), Myanmar, Singapore, Thailand and Vietnam. Also found in China and India.

HABITS AND HABITAT Nocturnal and strongly arboreal, this civet is usually found alone or occasionally in pairs. It is primarily frugivorous, although a variety of small animals are sometimes included in the diet. It is found in primary or tall secondary forests up to at least 1,200m.

NOTES The loud shrill call of this civet helps one locate this species in the high trees at night. There are a number of subspecies, one of which, the Javan *A. t. trilineata*, may be a distinct species, but more research is required before this can be confirmed.

Banded Linsang ■ *Prionodon linsang* HB 35–45cm, T 30–42cm

DESCRIPTION A small, long-bodied carnivore with buffy to golden fur and bold dark brown to black spots, which form approximately five bands across the back. The sides and legs are spotted. The long tail is banded with about seven broad dark rings. Spots on the side of the neck form longitudinal lines. The head is narrow with a pointed muzzle. It has short legs and, unlike civets, retractile claws.

DISTRIBUTION Brunei, Indonesia (Java, Kalimantan, Sumatra), Malaysia (Peninsular Malaysia, Sabah, Sarawak), Myanmar and Thailand.

HABITS AND HABITAT Little is known about the ecology of this nocturnal species. Found in primary and secondary forests, forest edges and disturbed forests, sometimes near human inhabited areas, it is largely arboreal but will forage on the ground as well, for birds, small mammals, reptiles and other small animals. It has been found up to 2,400m.

NOTES The two species of Asian linsang were until recently considered to be in the Viverridae family, with civets, but are now considered to belong to a family of their own (Prionodontidae).

Javan Mongoose ■ *Urva javanica* HB 36–42cm, T 27–32cm

DESCRIPTION The Javan Mongoose is a relatively small mongoose, with the females considerably smaller than the males. Speckled brown to reddish fur overall, with the head often especially reddish. Muzzle pointed, with a light brown to pinkish bare nose. Ears are small and rounded. Legs are relatively short, with small feet, and the tail is somewhat bushy and long.

DISTRIBUTION Cambodia, Indonesia (Java, Sumatra), Laos, Malaysia (Peninsular Malaysia), Myanmar, Thailand and Vietnam. Also China. Formerly confused with *Urva auropunctata*, which is in Myanmar as well as Afghanistan, Bangladesh, Bhutan, China, India, Nepal and Pakistan.

HABITS AND HABITAT Found in a variety of habitats, especially in well-watered, naturally open forests and grasslands. It appears to prefer edge habitat in most areas and can be found in disturbed forests and scrubland, often close to human-inhabited areas. Like all Asian mongoose species, it is terrestrial and very rarely climbs trees. Largely diurnal, though occasionally active at night, hunting birds, rodents, reptiles, amphibians and other small animals.

NOTES Can sometimes be studied by the quiet observer, as it frequents open areas and forest edges often near human settlement, darting in and out of cover.

Collared Mongoose

■ *Urva semitorquata*
HB 40–46cm, T 26–30cm

DESCRIPTION It has a reddish-brown coat with fine yellow markings on its back. Its throat is buff-orange and the lower parts of its legs are blackish brown. Sometimes bright orange coat, inviting confusion with Malay Weasel *Mustela nudipes*.

DISTRIBUTION Brunei, Indonesia (Kalimantan, Sumatra) and Malaysia (Sabah, Sarawak), possibly Philippines (Palawan).

HABITS AND HABITAT The ecology of the Collared Mongoose is poorly known. It is thought to be found in a variety of habitats including tall and secondary forests, and in disturbed areas. It is mainly terrestrial and diurnal, and probably also active at night. Its diet includes small animals. Further studies are needed to determine its distribution and general life history.

Short-tailed Mongoose ■

Urva brachyura
HB 38–45cm, T 20–25cm

DESCRIPTION The coat is blackish brown, with orange speckling that is obvious only at close range, and a pale brown chin and throat. The head and tail are paler than the rest of the body.

DISTRIBUTION Brunei, Indonesia (Kalimantan, Sumatra), Malaysia and Philippines (Palawan and the Calamian Islands).

HABITS AND HABITAT Like many small carnivores in the region, little is known of its ecology, though is thought to be found near rivers, in lowland primary and secondary forest and in plantations. It is mainly diurnal and terrestrial, with a diet that includes small animals and arthropods.

Tiger ▪ *Panthera tigris* HB 1.7–2.3m, T 0.95–1.15m

DESCRIPTION Unmistakable, with its large size and dark stripes on a deep orange coat. It has a pale underside and a ringed tail.

DISTRIBUTION Cambodia, Indonesia (Sumatra), Laos, Malaysia (Peninsular Malaysia), Myanmar, Thailand and Vietnam. Also found in Bangladesh, Bhutan, China, India, Nepal and Russia. In Southeast Asia, extirpated in Bali, Java and Singapore.

HABITS AND HABITAT Found in a variety of habitat types, from lower montane forest to riverine woodlands and peat swamps. Preys mainly on deer and wild pig, but also opportunistically feeds on a variety of other species, including ungulates much larger than itself. Generally solitary.

NOTES The Tiger's stripe patterns are like human fingerprints, and are used by researchers to identify individuals.

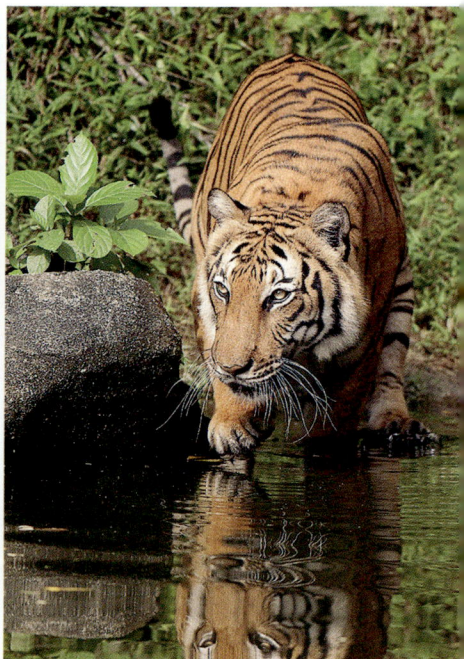

Leopard ▪ *Panthera pardus* HB 1–1.3m, T 0.8–1m

DESCRIPTION The Leopard is a large cat that occurs in two colour phases. The spotted form has black rosette spots on a yellowish-brown coat, and the melanistic form (often called the 'panther') has less obvious black rosette spots on a black coat, giving it an appearance of being an all-black cat, as the rosettes are only obvious in good light. The dark phase is more common in southern Thailand and Peninsular Malaysia. Nine subspecies are currently recognised globally.

DISTRIBUTION Cambodia, Indonesia (Java), Laos, Malaysia (Peninsular Malaysia), Myanmar, Thailand and Vietnam. A very widespread species, ranging throughout Africa and Central Asia to the Russian Far East.

HABITS AND HABITAT Found in all habitat types, Leopards have a very variable diet, taking a wider range of prey than any other cat. Frequently rests in trees.

Sunda Clouded Leopard ■ *Neofelis diardi* HB 61–106cm, T 55–91cm

DESCRIPTION Irregular cloud-shaped markings on a greyish-yellow coat. It has large, black ovals on its underbelly and limbs and there are two dark bars on the back of its neck. Long, thickly furred tail. Proportionately, it has the longest canine teeth of any living cat.

DISTRIBUTION Brunei, Indonesia (Kalimantan, Sumatra) and Malaysia (Sabah, Sarawak).

HABITS AND HABITAT Primarily nocturnal, with crepuscular activity peaks. Highly arboreal, well adapted to climbing and creeping in the trees with its stout legs and broad paws. Forest-dependent, but it is found, perhaps at lower density, in logged forest. Preys on a wide variety of species, ranging from deer, pigs, primates and other mammals to fish.

NOTES This was considered a subspecies of the Mainland Clouded Leopard (*N. nebulosa*) until research in 2006 determined that it is a distinct species. It is Borneo's largest cat.

Marbled Cat ■ *Pardofelis marmorata* HB 45–53cm, T 47–55cm

DESCRIPTION This cat has thick fur with a complex marbled pattern of dark splotches and black lines on its back and stripes on its head. It has a very long thick tail.

DISTRIBUTION Brunei, Cambodia, Indonesia (Kalimantan, Sumatra), Laos, Malaysia (Peninsular Malaysia, Sabah, Sarawak), Myanmar, Thailand and Vietnam. Also found in Bhutan, China, India and Nepal.

HABITS AND HABITAT Primarily forest-dependent. Preys on small animals including rodents.

NOTES Studies show that the Marbled Cat is very closely related to the big cats, possibly similar to big cat ancestors 10 million years ago, but may have diminished in size recently due to competition.

Asian Golden Cat ■ *Catopuma temminckii* HB 76–84cm, T 43–50cm

DESCRIPTION Coat varies from golden brown to tawny brown to greyish or black. Occasionally spotted. Black and white stripes on sides and front of head.

DISTRIBUTION Cambodia, Indonesia (Sumatra), Laos, Malaysia (Peninsular Malaysia), Myanmar, Thailand and Vietnam. Also found in Bangladesh, Bhutan, China, India and Nepal.

HABITS AND HABITAT Found in various forest habitats at a wide range of altitudes. Not primarily nocturnal, as previously thought. Mainly terrestrial but can climb trees.

NOTES On Borneo, the species is replaced by the closely similar Bay Cat *Catopuma badia*.

Flat-headed Cat ▪ *Prionailurus planiceps* HB 44–51cm, T 13–17cm

DESCRIPTION Narrow, flattened head with small ears. Semi-webbed toes. Greyish-brown coat with fine black speckling. Its chin and chest is white and it has faint pale and dark stripes on the sides of its face and forehead.

DISTRIBUTION Brunei, Indonesia (Kalimantan, Sumatra), Malaysia (Peninsular Malaysia, Sabah, Sarawak), and Thailand.

HABITS AND HABITAT Nocturnal and terrestrial, and found mainly in tall lowland forests, preferring areas close to streams. It is semi-aquatic, and its main diet is probably fish.

Mainland Leopard Cat ■ *Prionailurus bengalensis* HB 40–55cm, T 23–29cm

DESCRIPTION This is a small, lean cat. It has black spots on a reddish-orange or yellowish-buff coat, and the size and pattern of the spots can vary between individuals. Spots may be rounded or elongated and often almost join to look like stripes on its back. It is often confused with the larger Fishing Cat (see p. 87).

DISTRIBUTION Cambodia, Laos, Malaysia (Peninsular Malaysia), Myanmar, Singapore, Thailand and Vietnam. Also found in Afghanistan, Bangladesh, Bhutan, China, Hong Kong, India, Japan (Nansei-shoto), North and South Korea, Nepal, Pakistan, Russia and Taiwan. Populations in Brunei, Indonesia (Java, Kalimantan, Sumatra), Philippines and Malaysia (Sabah, Sarawak) now considered a separate species, Sunda Leopard Cat, *P. javanensis*.

HABITS AND HABITAT Relatively common, widespread in variety of habitats. Common in secondary forest and also in agricultural plantations. Preys mainly on small mammals. Nocturnal and mainly terrestrial, though it can climb small trees and is also a good swimmer.

Fishing Cat ■ *Prionailurus viverrinus* HB 72–78cm, T 25–29cm

DESCRIPTION A medium-large cat with light grey or olive-brown fur. It has black spots on its head and neck, small spots in rows on its flanks and back and a paler underside. It has partially webbed feet and a short tail.

DISTRIBUTION Cambodia, Indonesia (Java), Myanmar, Thailand and Vietnam. Presence in Laos, Peninsular Malaysia and Sumatra uncertain. Also found in Bangladesh, Bhutan, India, Nepal and Sri Lanka. Possibly extirpated in Pakistan.

HABITS AND HABITAT Swamps, marshy areas, oxbow lakes, mangroves. Widely distributed but highly localised. At home in the water, it feeds mainly on fish, crabs and molluscs.

> **CETACEANS**
> Whales, dolphins and porpoises are cetaceans. Cetaceans are broadly split into two groups: the large baleen whales such as the Blue Whale (*Balaenoptera musculus*), and the toothed whales, such as the Sperm Whale (*Physeter macrocephalus*), along with all dolphins and porpoises.

Pantropical Spotted Dolphin ■ *Stenella attenuata* TL up to 2.6m

DESCRIPTION The Pantropical Spotted Dolphin is dark grey with a paler underside, usually heavily spotted. Slender elongated body with a long narrow beak, tipped with white. A dark stripe runs from beak to flipper. Prominent tall dorsal fin.

DISTRIBUTION Cambodia, Indonesia, Malaysia, Myanmar, Philippines, Singapore, Thailand and Vietnam. Presence in Brunei uncertain. Inhabits tropical, equatorial and southern subtropical water bodies worldwide. Most abundant near the equator.

HABITS AND HABITAT Off shore, spotted dolphins feed largely on small fish, squid and crustaceans. In some areas, flying fish are also important prey. The diet of coastal populations is poorly known, but is thought to consist mainly of larger fishes, perhaps mainly bottom-living species. Known for their impressive high leaps into the air.

NOTES This species is often associated with schools of tuna, and it has therefore been heavily impacted in the past by accidental killing in nets. Some deliberate hunting occurs in the region, including hand-harpoon fisheries in the Philippines.

Spinner Dolphin ■ *Stenella longirostris* TL 1.3–2.3m

DESCRIPTION Three-part colour pattern, dark on top, lighter grey on the sides and pale underparts. Long slender beak, with the upper jaw dark grey and the lower jaw cream. Dark stripe from eye to flipper. Gently sloping melon and a prominent dorsal fin.

DISTRIBUTION Brunei, Cambodia, Indonesia, Malaysia, Myanmar, Philippines, Singapore, Thailand and Vietnam. Found in tropical and subtropical waters worldwide. Four subspecies are recognised. In Southeast Asia the Dwarf Spinner Dolphin (*S. l. roseiventris*) is distributed in shallow waters of inner Southeast Asia, including the Gulf of Thailand, the Timor and Arafura Seas off northern Australia, and other similar shallow waters off Indonesia and Malaysia. It is replaced in deeper and outer waters by the larger pelagic subspecies *S. l. longirostris*.

HABITS AND HABITAT In most tropical waters, nearly all records of Spinner Dolphins are associated with inshore waters, islands or banks, although sometimes in very large numbers hundreds of kilometres from the nearest land. The Dwarf Spinner Dolphin in Southeast Asian waters apparently inhabits shallow coral-reef habitat. Known for its incredibly high, spinning leaps. Often found in close association with Pantropical Spotted Dolphin (see p. 88), Yellowfin Tuna (*Thunnus albacares*) and birds of several species.

NOTES Throughout their range, Spinner Dolphins are taken as by-catch in purse-seine, gillnet and trawl fisheries, and some populations have been reduced by more than half.

Irrawaddy Dolphin ■ *Orcaella brevirostris* TL 2–2.8m

DESCRIPTION Grey to light grey, underparts paler. A very small dorsal fin behind the middle of the back. A high melon, rounded head and no beak. Large, rounded flippers.

DISTRIBUTION Brunei, Cambodia, Indonesia, Laos, Malaysia, Myanmar, Philippines, Singapore, Thailand and Vietnam. Also found in Bangladesh and India.

HABITS AND HABITAT Mainly found in shallow coastal waters, at brackish river mouths, near mangroves and in some major river systems, sometimes entering tributary rivers and lakes. Freshwater subpopulations are found in the Irrawaddy (up to 1,400km upstream) in Myanmar, the Mahakam (up to 560km upstream) in Indonesia, and the Mekong (up to 690km upstream) in Cambodia, Laos and Vietnam. Rarely leaps, and is quite quiet by nature.

NOTES The Irrawaddy Dolphin is threatened largely by entanglement in fishing gear, collisions with boats and habitat degradation. Capture for aquarium display is also a threat. More research and conservation efforts are urgently required.

False Killer Whale ■ *Pseudorca crassidens* TL up to 6m

DESCRIPTION The False Killer Whale, like all dolphins and porpoises, is a toothed whale. Uniformly dark except for paler areas on the throat and chest. Long slender body with a slender rounded head, with no beak. In adult males, the melon overhangs the lower jaw. Prominent dorsal fin and uniquely obvious elbows in the fins.

DISTRIBUTION Brunei, Cambodia, Indonesia, Malaysia, Myanmar, Philippines, Singapore, Thailand and Vietnam. Found in tropical and temperate waters worldwide.

HABITS AND HABITAT Usually found in relatively deep, offshore tropical and subtropical warm waters. Occasionally found in shallow and higher-latitude waters. Preys primarily on fish and cephalopods, but has been known to attack small cetaceans.

NOTES Normally encountered far off shore. They are suspected to be vulnerable to loud anthropogenic sounds, such as those generated by navy sonar and seismic exploration.

Melon-headed Whale ■ *Peponocephala electra* TL 2.2–2.6m

DESCRIPTION Dark grey to black, with a darker dorsal cape and facial markings. Underparts grey to white. Slim pointed head with no beak. Light grey or white lips. Large falcate fin.

DISTRIBUTION Brunei, Cambodia, Indonesia, Malaysia, Myanmar, Philippines, Singapore, Thailand and Vietnam. Pantropical distribution.

HABITS AND HABITAT Found in deep tropical and subtropical waters worldwide. Rarely ventures close to shore, except where water is deep. Sometimes found in groups of several hundred. Feeds largely on fish, squid and some crustaceans.

NOTES Sometimes travels with Fraser's Dolphins (*Lagenodelphis hosei*).

Dugong ▪ *Dugong dugon* TL 2.5–3.3m

DESCRIPTION Dugongs are closely related to the manatees (*Trichechus* spp.) of America and Africa, looking fairly similar except that manatees have a round tail fluke with a single lobe and a divided upper lip. Dugongs are grey-brown in colour, with a long fusiform body, no dorsal fin, and whale-like tail flukes. A striking feature is the fleshy oral disk, an expanded region between the mouth and nose, which is covered with vibrissae. The location of the nostrils at the tip of the snout enables dugongs to breathe discreetly with only their nostrils out of the water. Dugongs have axillary mammary glands (teats) located under each flipper. Tusks erupt in adult males and a few very old females, but do not extend beyond the premaxilla. Weighs up to 500kg.

DISTRIBUTION Brunei, Cambodia, Indonesia, Malaysia, Myanmar, Philippines, Singapore, Thailand and Vietnam. Also found in tropical and subtropical coastal waters of more than 40 other countries.

HABITS AND HABITAT Usually found in small groups, but can be found in large herds in some parts of its range. The Dugong, a seagrass specialist, is the world's only herbivorous mammal that is strictly marine. Tides can restrict its foraging on intertidal seagrass meadows on a daily basis.

NOTES The Dugong makes bird-like chirps that are inaudible above water. It is the only surviving member of the family Dugongidae, and very likely one of the most threatened marine mammals in Southeast Asia, due to hunting, entanglement in fishing gear, collisions with boats and habitat destruction.

Asian Elephant ■ *Elephas maximus* SH 1.5–3m, HB 3–6m, T 1–1.5m

DESCRIPTION Greyish-brown thick and wrinkly skin. Not all elephants have tusks; some adult males have long tusks reaching up to 2m, but females and younger males have tushes, which are shorter tusks that are usually not visible. Its back is rounded and sloped, making its crown the highest point of its body. Asian Elephants have smaller ears than African elephants (*Loxodonta* spp.).

DISTRIBUTION Cambodia, Indonesia (Kalimantan, Sumatra), Laos, Malaysia (Peninsular Malaysia, Sabah), Myanmar, Thailand and Vietnam. Also found in Bangladesh, Bhutan, China, India, Nepal and Sri Lanka. Extirpated in Pakistan.

HABITS AND HABITAT A highly social animal, living in herds of related females and juveniles, led by a matriarch. Males leave their herds when they mature at about 6–7 years of age and become predominantly solitary, seeking females for mating purposes only. An elephant can consume up to 150kg of vegetation daily. It eats mainly grasses but also fruit, bark and other plants. It is an excellent swimmer. Elephants practise allomothering, where females in the herd care for and protect all the young in the herd as they would their own.

▪ ELEPHANT ▪

ABOVE AND BELOW, RIGHT: *Pygmy Elephant*

NOTES The elephants in
the Malaysian state of Sabah
are sometimes considered a
distinct subspecies, the Bornean
Elephant, also called the Pygmy
Elephant because of its smaller
size. This is still subject to debate,
however, and some believe that
these elephants, which have a
highly limited distribution, are
not native, and may actually be
descendants of an introduced
population. A good place to view
them is along the Kinabatangan
River, by boat.

Asian Tapir ■ *Tapirus indicus* SH 0.9–1.05m, HB 2–2.4m, T 5–10cm

DESCRIPTION The closest relatives of the tapirs are horses and rhinoceroses. There are four tapir species globally but only one occurs in Asia. The Asian Tapir is the largest of the tapirs, with a stocky build and very distinctive black and white colouring. The front part of the body, including the head and forelegs, and the hind legs, are black. The remainder of the body is white. Its elongated nose and upper lip form a prominent prehensile proboscis, an important adaptation that the tapir uses to grasp vegetation. Infants are born dark grey with short, horizontal white stripes and spots in rows.

HABITS AND HABITAT Found in primary and secondary forest, both montane and lowland. While it seems to prefer intact primary rainforest, it is also found in secondary growth and degraded habitat as well. Often prefers streams and rivers. Largely solitary by nature and mainly nocturnal. Eats a wide variety of plant matter.

DISTRIBUTION Indonesia (Sumatra), Malaysia (Peninsular Malaysia), Myanmar and Thailand.

NOTES At a glance, the black and white colouration may seem very conspicuous, but this disruptive patterning actually helps break up its body outline in shady and moonlit forests, just like the tiger's striped coat.

> **Rhinoceros**
> Globally there are five species of rhinoceros, and Asia is home to three while the other two are found in Africa.

Sumatran Rhinoceros ■ *Dicerorhinus sumatrensis*

SH 1.2–1.3m, HB 2.4–2.6m, T 35–70cm

DESCRIPTION The Sumatran Rhinoceros is dark brownish grey with a large heavy body and thick skin. One fold of skin crosses the back behind the shoulders. Sparse hair covers the body, especially in young animals. Three toes on each foot. Two horns, with the front horn longer and thinner than the much smaller rear horn, which is sometimes barely visible. This is the smallest of the rhinos.

DISTRIBUTION Indonesia (Kalimantan, Sumatra). Until recently found in Malaysia (Peninsular Malaysia, Sabah, Sarawak), Myanmar, Brunei, Cambodia, Laos, Thailand and Vietnam but now extirpated from all these areas. Formerly also occurred in Bangladesh, Bhutan and India, but also extirpated there.

HABITS AND HABITAT The Sumatran Rhinoceros is extremely shy, and is generally solitary except for mating pairs and mothers with young. It occurs from sea level to over 2,500m, inhabiting tropical rainforest and montane moss forest, and occasionally at forest margins and in secondary forest. Found mainly in hilly areas near water sources, spending the hotter parts of the day resting, often in wallows of mud.

NOTES The two principal threats are poaching for the horns and the resulting reduced population viability. The horns are believed to have medicinal properties, and despite trade in these being illegal, poaching continues, leaving the total population at an estimated 275 or fewer. It is of paramount importance that efforts to save this species from extinction are stepped up in effectiveness and scale.

Javan Rhinoceros ■ *Rhinoceros sondaicus* SH 1.6–1.8m, HB 3–3.2m, T 70cm

DESCRIPTION Large and heavy with thick dark grey skin, with three folds of skin across the back. A single horn at the tip of the snout, often inconspicuous in females.

DISTRIBUTION Indonesia (Java). Extirpated in Cambodia, Indonesia (Sumatra), Laos, Malaysia (Peninsular Malaysia), Myanmar, Thailand and Vietnam. Also extirpated in Bangladesh, China and India. Originally three recognised subspecies: *R. s. sondaicus*, *R. s. annamiticus* and *R. s. inermis* (the latter two extinct).

HABITS AND HABITAT This is a lowland species that typically occurs up to 600m, but has been recorded above 1,000m. Formerly occurred in more open mixed forest and grassland and on high mountains. Because of its extreme rarity, little is known about its preferred habitat.

NOTES Poaching for the horn has pushed the Javan Rhinoceros to the brink of extinction. An estimated 40–60 animals live in the area on the western tip of Java in Ujung Kulon National Park, and nowhere else. Failure to protect this species in Vietnam, the last population outside of Java, led to its extirpation there in 2011. Efforts to save this species from total extinction are needed urgently.

Eurasian Wild Pig ▪ *Sus scrofa* SH 60–80cm, HB 135–150cm, T 20–30cm

DESCRIPTION The colour of this species varies from reddish to blackish to grey, with long black hairs on its upper back and neck, forming a mane. It has an elongated muzzle, with males having enlarged, protruding canines. Juveniles are dark brown with elongated white stripes along the body. It is similar to the Sunda Bearded Pig (see p. 101), but is smaller and lacks the extensive beard.

DISTRIBUTION Cambodia, Indonesia, Laos, Malaysia (Peninsular Malaysia), Myanmar, Singapore, Thailand and Vietnam. The Eurasian Wild Pig has the largest range of all pigs, found in numerous countries worldwide including much of Europe, North Africa and mainland Asia.

HABITS AND HABITAT Found in a wide variety of habitats, from mature and secondary forests to disturbed areas and plantations. It is an omnivore, eating roots, tubers, fruit, seeds, other vegetation and animal matter encountered on the ground, such as eggs, nestlings and worms. Mainly active early in the day and late afternoon, but human disturbance can make them nocturnal. Lives in large herds of up to 20 individuals, though there have been instances of over 100 gathering.

NOTES At a global level, there are no major threats. However, there are many threats at a local level, principally hunting pressure, for food, for sport or in reprisal for crop damage.

Javan Warty Pig ■ *Sus verrucosus* SH 70–90cm, HB 90–190cm

DESCRIPTION Reddish in colour, although sometimes appearing blackish. Underparts are white or yellowish. Large head, with large ears, and in males three pairs of facial 'warts', lacking in females. Males are up to twice as large as females. Both sexes have a long mane, often a lighter orange-brown colour, on top of the head and along the spine to the rump. Long thin legs and a long tail with a tuft of hair at the end. Two subspecies are recognised. In the field, they can appear similar to Eurasian Wild Pig (see p. 99).

DISTRIBUTION This species is endemic to Indonesia, where it was historically found on Java, Madura Island (extirpated) and Bawean Island. It now remains in only small pockets of fragmented and rapidly shrinking habitat – on Java it now survives in at least 10 separate, isolated areas.

HABITS AND HABITAT Occurs in cultivated areas and teak plantations, as well as in remaining stands of forest. It appears to thrive in mosaics of forest–teak plantations, scrub and open grasslands, and seems to prefer secondary forests over primary forests. It is restricted to elevations below 800m and is often found in mangrove and swamp forests. Lives in small groups of 4–6 individuals.

NOTES This species is in trouble. Its decline is due to habitat fragmentation, hybridisation with Eurasian Wild Pig and intense hunting pressure, for food, for sport and to protect crops. This species is also (illegally) pitted against dogs for sport. Captive breeding programmes are under way to ensure that the Javan Warty Pig is not lost.

Sunda Bearded Pig ▪ *Sus barbatus* SH 70–90cm, HB 120–150cm, T 17–25cm

DESCRIPTION Colour varies from blackish (in young animals) to grey, reddish brown or yellowish brown. Appearance is affected by the colour of mud the pig has been wallowing in. Large, long head with long thick bristles on snout. A fleshy protuberance on the sides of the snout above the mouth. Lower canines of males protrude. Hoof prints are rounded and symmetrical, with distinct dew toes.

DISTRIBUTION Brunei, Indonesia (Kalimantan, Sumatra), Malaysia (Peninsular Malaysia, Sabah, Sarawak) and Philippines (extreme south).

HABITS AND HABITAT Diurnal and nocturnal – largely nocturnal in areas with heavy hunting pressure. Most often found in tropical evergreen rainforest, but also in a wide variety of habitat types, ranging from beaches to upper montane forests. Large-scale population movements, over weeks or months, have often been recorded, reportedly linked to the availability of seasonal fruits, particularly mast-fruiting species, i.e. those with huge variation between years in fruit production.

NOTES Borneo represents the best place to view this species, although an introduced disease, African swine fever, has caused major declines in many populations. They may be seen with a great deal of luck in other parts of their range, such as in Endau Rompin National Park in Peninsular Malaysia.

> **CHEVROTAINS**
> The chevrotains are also known as mouse-deer, referring to their small size – though they are not deer.

Lesser Oriental Chevrotain ▪ *Tragulus kanchil*

SH 20–23cm, HB 40–55cm, T 6–9cm

DESCRIPTION The upperparts are reddish brown mixed with fine black fur, the centre of the nape being darker than the rest of the back, often appearing like a dark stripe. It has white underparts, with variable brown stripes in the middle and along the sides of its body. It has distinctive dark brown and white markings on its throat and upper chest, typically with a triangular white stripe in the middle, bordered by a dark brown triangle, with diagonal white stripes on the side, which usually join at the chin. Legs very slender. Males have visible protruding canines.

DISTRIBUTION Brunei, Cambodia, Indonesia, Laos, Malaysia (Peninsular Malaysia, Sabah, Sarawak), Myanmar, Singapore, Thailand and Vietnam. Presence in China uncertain.

HABITS AND HABITAT Lowland primary and secondary forest, as well as cultivated areas. It is usually solitary, and is active periodically during both night and day. Its diet comprises shoots, young leaves, fallen fruit and fungi.

Greater Oriental Chevrotain ▪ *Tragulus napu*

SH 30–35cm, HB 52–57cm, T 6–10cm

DESCRIPTION This species has coarsely mottled orange-buff, grey-buff and blackish upperparts, which are darker in the midline and paler along the sides of its body. Often, it has a darker nape patch. The intensity of the colouration varies between individuals. It has white underparts, usually without belly stripes. It has a pattern of brown and white markings on the underside of its neck and upper chest, typically with a triangular white stripe in the centre bordered by dark brown stripes. It has two white stripes on each side, which appear as two separate white bars on the side of the neck when viewed in profile.

DISTRIBUTION Brunei, Indonesia (Kalimantan, Sumatra), Malaysia (Peninsular Malaysia, Sabah, Sarawak), Myanmar, Singapore and Thailand.

HABITS AND HABITAT Found in mainly tall and secondary forest. It is mainly nocturnal but can be active during the day as well. Its diet includes leaf shoots, fallen fruit and other vegetation.

NOTES The Greater Oriental Chevrotain was thought to be extirpated in Singapore until it was rediscovered in 2008 on Pulau Ubin, a small island off the northeastern corner of mainland Singapore.

Balabac Chevrotain ■ *Tragulus nigricans* SH 18cm, HB 40–50cm, T 8cm

DESCRIPTION Upperparts washed with black, mottled with orange, with three narrow white bars on the throat and chest, starting from a white patch under the chin. Its nose bridge and forehead are dark brown, leading to a dark crown. Like other chevrotains, it is small, with very slender legs.

DISTRIBUTION Philippines (Balabac, Bugsuc and Ramos Islands). Its presence in Malaysia is uncertain, making it endemic to the Philippines until confirmed otherwise.

HABITS AND HABITAT Not much is known about this species, but it is known to occur in primary and secondary lowland forest and shrubland, and it may frequent mangroves and more open areas to forage.

NOTES The chevrotains on the small Malaysian island of Pulau Banggi, located midway between Balabac and the Bornean mainland, might belong to this species – but further studies are required.

Red Muntjac ▪ *Muntiacus muntjak* SH 50–55cm, HB 90–110cm, T 17–19cm

DESCRIPTION A small deer with reddish-brown to paler reddish-yellow coat, generally darker along the midline, with paler underparts. The underside of the tail is white. Males have small, thick antlers with a small spike at the base and the pedicels have very obvious black lines, which continue onto the antlers. Females have a stiff tuft of hair on the top of the head instead of antlers and pedicels. Juveniles have white spots, which they lose as they mature.

DISTRIBUTION Brunei, Cambodia, Indonesia (Java, Kalimantan, Sumatra), Laos, Malaysia (Peninsular Malaysia, Sabah, Sarawak), Myanmar, Thailand and Vietnam. Extirpated in Singapore. Also found in Bangladesh, Bhutan, China, Hong Kong, India, Nepal, Pakistan and Sri Lanka.

HABITS AND HABITAT Found in a variety of habitats, from tropical to dry dipterocarp, lowland and hill forests. It mainly eats leaves, shoots and fallen fruit. Primarily nocturnal, although in sites with less hunting and other human disturbance they are also active during the day.

NOTES Both males and females emit an alarm call that sounds like a bark, which gives this species its other common name, the Common Barking Deer.

Sambar ■ *Rusa unicolor* SH 140–160cm, HB 150–200cm, T 21–28cm

DESCRIPTION A large-bodied deer. Dark to greyish brown, darker along the midline. Adult males have three tines to each antler, and long, coarse neck hair. The young may have light spots.

DISTRIBUTION Brunei, Cambodia, Indonesia (Kalimantan, Sumatra and some smaller islands), Laos, Malaysia (Peninsular Malaysia, Sabah, Sarawak), Myanmar, Thailand and Vietnam. Also found in Bangladesh, Bhutan, China, India, Nepal, Sri Lanka and Taiwan.

HABITS AND HABITAT Mainly nocturnal. Eats grasses, herbs, shrubs, leaves. Frequents salt licks. Usually solitary and nocturnal, possibly a consequence of severe hunting pressure. Very much declined in many areas due to poaching.

NOTES Easily viewed in certain protected areas in Thailand, such as Khao Yai National Park, Thap Lan National Park and Phu Khieo Wildlife Sanctuary.

Eld's Deer ■ *Rucervus eldii* SH 120–130cm, HB 150–170cm, T 22–25cm

DESCRIPTION Brown with white belly. The adult male's antlers have a prominent brow tine that forms a continuous curve with the main branch, which has a few small tines at its tip. The bow-shaped antlers grow outwards and then inwards, rather than just upwards.

DISTRIBUTION Cambodia, Laos and Myanmar. Possibly extirpated in Thailand and Vietnam. Also found in China and India.

HABITS AND HABITAT Found in lowland swamps, dry dipterocarp and grasslands. Diet of grass, browse, fallen fruit and flowers. Forms large herds where not hunted. Due largely to poaching, this species has undergone severe population declines.

MAIN PICTURE: *Male.* INSET: *Female with fawn*

Hog Deer ■ *Axis porcinus* SH 65–72cm, HB 140–150cm, T 17–21cm

DESCRIPTION Similar to Sambar (see p. 106), but smaller, with shorter legs. Light brown fur. Slender antlers on males. Young are heavily spotted. Females may retain spots.

DISTRIBUTION Cambodia, Myanmar, and reintroduced in Thailand. Also found in Bangladesh, Bhutan, India, Nepal and Pakistan. Presence in Laos and Vietnam uncertain. Extirpated in China.

HABITS AND HABITAT Seasonally inundated (natural flooding) lowland grasslands. Mainly feeds on grasses.

NOTES Efforts to reintroduce the species in Thailand are ongoing. It has been extirpated mainly because of over-hunting and habitat loss. Its main habitat is prime rice growing land – rice is a grass that grows best under seasonal inundation. So it remains only in small isolated areas from which it is easily hunted out.

Calamian Deer ▪ *Axis calamianensis* SH 60–100cm, HB 100–175cm, T 12–38cm

DESCRIPTION Heavy-bodied, though fairly small. It is tawny-brown with darker legs and underparts. It has subtle white markings around its muzzle, and the underside of the tail is also white. Males have three-pronged antlers.

DISTRIBUTION Philippines. Endemic to the Calamian islands of the Palawan Faunal Region.

HABITS AND HABITAT Found in grasslands, open woodlands and secondary forest. It is diurnal, and eats mainly leaves. It lives in small herds, generally between 7 and 14 individuals, but groups of up to 27 have been recorded. In heavily hunted areas, groups tend to be smaller.

> **CATTLE, BUFFALO, GOATS AND SHEEP**
> Cattle, buffalo, goats and sheep belong to the family Bovidae, which also includes
> antelopes. One of the defining characteristics of this family is the presence of
> unbranched horns, which are never shed and continue to grow as the animal ages.

Takin ▪ *Budorcas taxicolor* SH 0.7–1.4m, HB 1–1.4m, T 7–12cm

DESCRIPTION The Takin has long and shaggy hair, varying in colour from dark brown to
a light golden brown. Females and young are often greyer than males. it is a thick-bodied
animal with an arched back, thick neck, thick legs and an outwardly curved face. Horns arise
from the middle of the forehead, turning outwards and then curving backwards. Females and
juveniles have straighter horns than males.

DISTRIBUTION Myanmar. Also found in Bhutan, China and India.

HABITS AND HABITAT Found in hill forests, feeding in open areas and sheltering in forests.
In summer, feeds in alpine meadows up to 4,000m, and in winter, in valleys and forests to
as low as 1,000m. Eats a variety of grasses, bamboo shoots, and leaves of shrubs and trees.
Mostly active in early morning and late afternoon. Often in small herds of 20–30, but
occasionally up to 300. Older males are solitary for much of the year.

NOTES Four subspecies are recognised, with only *B. t. taxicolor* found in Southeast Asia, in
northern Myanmar.

Banteng ■ *Bos javanicus* SH 1.5–1.7m, HB 1.9–2.25m, T 65–70cm

DESCRIPTION Males are dark brown, often black, while females and immature males are bright rufous brown. Both sexes have a white band across the muzzle, white buttocks and white 'stockings' on the lower parts of the legs. Mature females are smaller than mature males. Horns curve outwards and forward, with a horny patch of skin between the horns.

DISTRIBUTION Cambodia, Indonesia (Bali, Java, Kalimantan), Laos, Malaysia (Peninsular Malaysia, Sabah, Sarawak), Myanmar, Thailand and Vietnam. Extirpated in Brunei. Also possibly found in southern China. Extirpated in Bangladesh and India.

HABITS AND HABITAT Prefers open flat or undulating terrain with dry deciduous forests and open grassland mosaics over closed evergreen forests, although in some areas it occupies secondary forest formations resulting from logging and fires. Due to human pressure, Banteng increasingly use less preferred habitat, and are somewhat adaptable. Feeding grounds tend to be near permanent water supplies.

NOTES Banteng are largely naturally diurnal, but hunting pressure has resulted in most populations adapting to a nocturnal existence. The largest remaining natural populations are in Indonesia (Java) and Thailand.

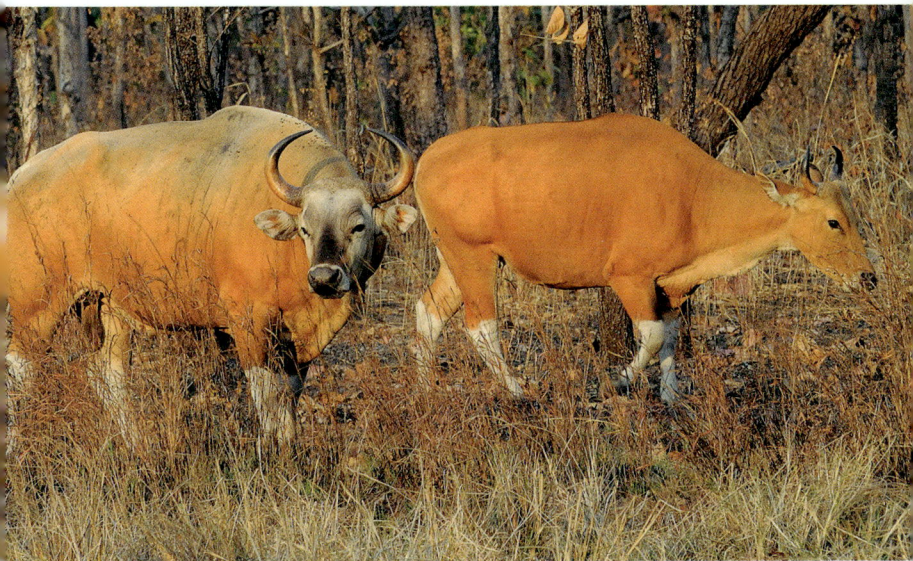

Gaur ▪ *Bos gaurus* SH 1.7–1.85m, HB 2.5–3m, T 70–105cm

DESCRIPTION Massive. Both sexes are very dark brown, almost black, with whitish or yellowish 'stockings' on the lower parts of the legs. Adult males have a thick muscular ridge on the back. Horns curve outwards and upwards. Lacks the white buttocks of Banteng.

DISTRIBUTION Cambodia, Laos, Malaysia (Peninsular Malaysia), Myanmar, Thailand and Vietnam. Also found in Bangladesh, Bhutan, China, India and Nepal.

HABITS AND HABITAT Forested areas, feeding on grasslands. Mainly nocturnal, and favours salt licks.

NOTES Heavily hunted in much of its natural range. Several protected areas in Thailand have growing populations, including Khao Yai National Park and Huai Kha Khaeng Wildlife Sanctuary.

Wild Water Buffalo ▪ *Bubalus arnee* SH 1.6–1.9m, HB 2.4–2.8m, T 60–85cm

DESCRIPTION Larger than Domestic Water Buffalo (*Bubalus bubalis*), with broader and wider-spread horns, the broadest of any living bovid.

DISTRIBUTION Cambodia, Myanmar and Thailand. Possibly extirpated in Vietnam. Extirpated in Laos. Also found in Bangladesh, Bhutan, India and Nepal. Possibly extirpated in Sri Lanka.

HABITS AND HABITAT Prefers grasslands and open forests, near water. Frequently wallows in mud and pools. Lives in small herds.

NOTES Fewer than 4,000 remain, with populations highly fragmented and threatened by hunting, habitat loss and interbreeding with Domestic Water Buffalo.

Tamaraw ■ *Bubalus mindorensis* SH 1–1.05m, HB 2.2m, T 60cm

DESCRIPTION A small buffalo, very dark brown in colour, with a white patch on the throat and sometimes white 'stockings' on the lower legs. Both sexes have horns that point backwards, and not in a broad arc as in the Domestic Water Buffalo (*Bubalus bubalis*). Males have larger horns than females.

DISTRIBUTION Philippines. Endemic to the island of Mindoro.

HABITS AND HABITAT Once widespread across Mindoro, in a range of habitats from sea level to mountainous areas, in marshes, bamboo forests, mixed forests and grasslands, this highly threatened species is now confined to two or three areas, in rough terrain in remote inaccessible areas. Mixed grassland and forest mosaics are most likely the preferred habitat. Largely solitary or in female–offspring pairs. Now extremely rare due to overhunting.

NOTES Because of hunting pressure, this formerly diurnal species has become nocturnal.

Maned Serow ▪ *Capricornis sumatraensis*

SH 85–94cm, HB 140–155cm, T 11–16cm

DESCRIPTION Black upper- and underparts. Hairs do not have white bases except those on the mane and along the back either side of black dorsal stripe. The mane varies from mostly white, or golden buffy-white, to black with only a few white hairs. Short white jaw streaks quickly becoming red then black. Long legs black, with distinct (varying) reddish tones towards the hoofs, occasionally with white hairs. There is a large open gland in front of each eye. The horns in both sexes curve back slightly from the forehead, with horizontal ridges.

DISTRIBUTION Cambodia, Indonesia (Sumatra), Malaysia (Peninsular Malaysia), Myanmar, Laos and Vietnam.

HABITS AND HABITAT This serow is often found on steep mountain slopes between 200 and 3,000m, covered by both primary and secondary forests. It is also often found on forested limestone karsts. Regular latrines are used, often under overhanging rocks or cliff faces. Solitary, but also recorded in small groups of up to seven.

NOTES Habitat loss, including limestone and quartz-ridge quarrying, as well as hunting, are serious threats. Formerly split into several species, but now considered only one.

Burmese Red Serow ▪ *Capricornis rubidus*

SH 85–95cm, HB 140–155cm, T 11–16cm

DESCRIPTION Red-brown, with more red on the neck and flanks, with hairs having black bases. The underside is white. There is a black dorsal line and a dark red mane, which is shorter than in other serow species, and a short tail. Nose, jaw and throat patch are white, creamy red or red. Long ears and a large open gland in front of each eye. Horns in both sexes curve back slightly from the forehead, with horizontal ridges.

DISTRIBUTION Myanmar. Also Assam and India, although some authorities consider those populations to be a different species.

HABITS AND HABITAT Tropical and subtropical hill and montane forests, including rugged limestone hills. Serow tend to be somewhat solitary, except females with young.

Saola ▪ *Pseudoryx nghetinhensis* SH 80–90cm, HB 150cm, T 23cm

DESCRIPTION Overall deep chestnut brown, varying from rich reddish brown to almost black. The anal area and inner flanks are white. A thin black stripe runs along the spine from between the shoulders to the top of the tail. A white horizontal stripe runs across the rump, with a white band across the tail. The legs are darker than the main body and have two white spots above the hooves. The face is marked with white to buff patches, the most distinctive of which is a long, thin 'eyebrow' stripe above each eye. There is a variable pattern of spots and slashes from beneath the eye to under the jaw, while a single white spot may be present on the cheek. White lips, underside of chin and upper throat. Both sexes have long, slender horns that curve slightly backwards and reach 35–50cm long.

DISTRIBUTION Laos and Vietnam.

HABITS AND HABITAT Found in forested habitats in the Annamite mountains, in closed canopy, broadleaf evergreen forests, usually at 400–800m above sea level. It has been suggested that there are seasonal movements between elevations, and different forest types. Forest blocks smaller than 25km^2 and agricultural areas are not used, highlighting the need to preserve large blocks of tall forest. Probably largely solitary, though sometimes in groups of two or three, and perhaps up to six or seven. Natural predators potentially include Leopard *Panthera pardus*, Tiger *P. tigris* and Dhole *Cuon alpinus*.

NOTES The Saola represents one of the more amazing animal discoveries in recent times, with the first record of this species coming to the scientific world only in 1992. Hunting, often with dogs, and death in snares that are often set for other species, represent the greatest threats to its survival, followed closely by habitat loss. Probably numbering less than a few hundred, it is one of the most seriously threatened large mammals in the world.

Black Giant Squirrel ■ *Ratufa bicolor* HB 37–41cm, T 42–50cm

DESCRIPTION Squirrels are small to large-sized rodents. The Black Giant Squirrel is the largest squirrel in Southeast Asia and quite variable in appearance. Upperparts, including the top of the head, are generally black, but can be brown or slightly reddish-tinged black. Underparts are pale cream or buff to orange, running down the front legs. The feet are blackish, as is the entire tail. The face is largely white to reddish with a black moustache-like stripe extending across the cheeks. One subspecies, *R. b. smithi*, is buff in colour from the nape, down the back and sometimes onto the tail, and has pale areas on front and hind limbs.

DISTRIBUTION Cambodia, Indonesia (Bali, Java, Sumatra), Laos, Malaysia (Peninsular Malaysia), Myanmar, Thailand and Vietnam. Also found in Bangladesh, Bhutan, China and Nepal.

HABITS AND HABITAT Diurnal and arboreal, occasionally venturing to forest floor to feed. Found in tropical and subtropical montane evergreen and dry deciduous forests, often in tall secondary forests, and in some parts of its range near human settlements. Can be quite common in areas with suitable habitat and low levels of hunting.

NOTES Although still fairly widespread, this species has an increasingly patchy distribution throughout its range, although possibly less so in Malaysia.

Cream-coloured Giant Squirrel ▪ *Ratufa affinis*

HB 31–38cm, T 37–44cm

DESCRIPTION Large. Buff-brown to orange-brown, becoming darker in colour in the northern parts of its range. Animals on Borneo (pictured) are much darker. Underparts are pale buff to whitish.

DISTRIBUTION Brunei, Indonesia (Kalimantan, Sumatra), Malaysia, (Peninsular Malaysia, Sabah, Sarawak), Singapore and Thailand. This species has not been seen in Singapore since 1995 and may be extirpated, due largely to habitat destruction.

HABITS AND HABITAT Diurnal and strictly arboreal, found in tall closed-canopy forests. It occurs in lowland and hilly areas, sometimes in selectively logged areas, but appears generally to avoid plantations.

NOTES This squirrel occurs in relatively low densities throughout its range, possibly because of competition for food with other arboreal vertebrates such as birds and primates. Loss of tall trees, as well as hunting for consumption, is a threat to this species in some areas.

Plantain Squirrel ▪ *Callosciurus notatus* HB 17–22cm, T 16–21cm

DESCRIPTION A medium-sized squirrel with brownish upperparts and underparts reddish orange, varying from pale to dark. The tail is tipped with orange, which is more obvious in some individuals than others. Two stripes run down its sides, with a dark stripe next to the belly and a light stripe above that.

DISTRIBUTION Brunei, Indonesia (Java, Kalimantan, Sumatra), Malaysia (Peninsular Malaysia, Sabah, Sarawak), Singapore and Thailand. Also found on smaller islands within this range.

HABITS AND HABITAT Diurnal and arboreal, this squirrel is quite adaptable, found in a wide variety of habitats, including secondary forests, plantations, parks, gardens and along forest edge. Less common in primary forests. Found from lowland up to 1,500m. Its diet consists mostly of fruit and bark, as well as some insects.

NOTES As it is very adaptable, its range has expanded with human-related habitat alteration. It is not currently threatened, although in some areas it is heavily hunted for consumption.

Slender Squirrel ■ *Sundasciurus tenuis* HB 11–16cm, T 12–13cm

DESCRIPTION Small, with slightly speckled brown-olive upperparts and light grey to whitish underparts. Slender tail, pale eye-rings, and a nose that appears slightly upturned.

DISTRIBUTION Brunei, Indonesia (Kalimantan, Sumatra and some smaller islands), Malaysia (Peninsular Malaysia, Sabah, Sarawak), Singapore and southern peninsular Thailand.

HABITS AND HABITAT Diurnal and arboreal, it is found in primary and secondary forests, in both lowlands and mountains, and in parks and gardens.

NOTES Similar to Low's Squirrel (see p. 122), but has greyer underparts and a longer, more slender tail. Has adapted to city park life in parts of its range and can be quite easily observed in Singapore.

Low's Squirrel ■ *Sundasciurus lowii* HB 13–15cm, T 8–10cm

DESCRIPTION Brown to reddish brown, sometimes with slight speckles. The underside is buffy, sometimes with a reddish colour. Slight reddish eye-ring and a very bushy tail.

DISTRIBUTION Brunei, Indonesia (Kalimantan, Sumatra and a few smaller islands), Malaysia (Peninsular Malaysia, Sabah, Sarawak) and peninsular Thailand.

HABITS AND HABITAT Diurnal, found in small trees and sometimes seen foraging on the ground. This small squirrel is fond of secondary and disturbed forests, but occurs at relatively low densities, possibly due to competition for food from other tree-dwelling vertebrates, such as birds, primates and other squirrels. It is usually found below 900m, but on Borneo it occurs up to 1,400m.

Western Striped Squirrel ■ *Tamiops mcclellandii*

HB 11–12.5cm, T 11–14cm

DESCRIPTION Upperparts are mottled grey and brown with five dark stripes and four pale stripes running the length of the back, with the central stripes being the darkest. The pale stripes vary from white to cream or buff, with the outer pair wider and brighter than the inner pair. The outer pale stripes are continuous with white stripes on the cheeks. Underparts are pale buff to orangeish. White tufts on ears. Superficially resembles chipmunks (*Tamias* spp.) and is sometimes mistakenly referred to as such.

DISTRIBUTION Laos, Malaysia (Peninsular Malaysia), Myanmar, Thailand and Vietnam. Also found in Bhutan, China, India and Nepal.

HABITS AND HABITAT
A wide array of habitats, including primary and secondary forests, scrub and degraded forests and gardens, especially with fruit trees and coconut palms. Often found in hilly or mountainous areas in the Sundaic part of its range, usually above 700m. In northern Southeast Asia, it is common at lower elevations.

NOTES Further research is required, as the genus *Tamiops* is in need of taxonomic review and this species may represent a species complex (currently there are four species described). Species range limits remain poorly understood. This tiny squirrel very often feeds in the proximity of mixed bird flocks, possibly taking advantage of the extra eyes watching out for danger.

Prevost's Squirrel ▪ *Callosciurus prevostii* HB 20–27cm, T 20–27cm

DESCRIPTION One of the most colourful squirrels in the region, with pelage differing greatly in various parts of the region. All forms have dark reddish or orange underparts. The mainland Southeast Asia form has black upperparts with a white stripe on the flanks. The sides of the face are grey. There are a number of subspecies on Borneo, ranging from entirely black upperparts to grizzled olive-buff or grizzled brown, some with reddish shoulders or thighs, and tails ranging from black to grizzled brown or grey.

DISTRIBUTION Brunei, Indonesia (Kalimantan, Sumatra), Malaysia (Peninsular Malaysia, Sabah, Sarawak) and peninsular Thailand. It also occurs on many smaller islands throughout this range.

HABITS AND HABITAT Diurnal and arboreal, rarely venturing to the ground to forage or to cross gaps in the forest. It is found in lowland primary and secondary forests, but also uses oil-palm and coconut plantations that are adjacent to forests.

NOTES Habitat continues to be destroyed and fragmented, with much of the natural habitat of this species being replaced by plantations. In some parts of its range it is threatened by hunting for consumption and to be used as pets. More research on the impact of these threats is needed.

Pallas's Squirrel ■ *Callosciurus erythraeus* HB 20–26cm, T 19–25cm

DESCRIPTION Variable in colouration throughout the region. In most, upperparts are agouti; brownish to greyish white. In some populations, sometimes with a black stripe down the back. Underparts are reddish or reddish brown to grey-brown. In some populations there is an agouti-coloured stripe running down the middle of the belly, separating the reddish colouration into two stripes. In some populations, the entire body is reddish, with underparts pale orange. The black dorsal stripe varies in width from one population to another, or is absent.

DISTRIBUTION Cambodia, Laos, Malaysia (Peninsular Malaysia), Myanmar, Thailand and Vietnam. Also found in Bangladesh, China, India and Taiwan.

HABITS AND HABITAT Diurnal and arboreal. Found in primary and secondary forests, as well as in orchards and disturbed habitat. Often in hilly areas. Nests in tree-hollows in mid-canopy.

Black Flying Squirrel ■ *Aeromys tephromelas* HB 33–43cm, T 41–47cm

DESCRIPTION A large dark grey to blackish flying squirrel with a dark face and a long dark tail. Some light speckles on the upperparts. Underparts are similar but paler.

DISTRIBUTION Brunei, Indonesia (Kalimantan, Sumatra), Malaysia (Peninsular Malaysia, Sabah, Sarawak) and southern Thailand.

HABITS AND HABITAT Like other flying squirrels, this species is nocturnal and arboreal. It appears to be quite adaptable and is found in primary forests and tall secondary forests, as well as in tree plantations and in gardens with tall trees. Shelters in hollows in trees. It is usually found in hilly areas.

NOTES Like the colugos (see pp. 17–18), flying squirrels do not, strictly speaking, fly – they glide from one tree to another, using the patagium (a membrane on each side of the body joining fore and hind limbs) like a glider.

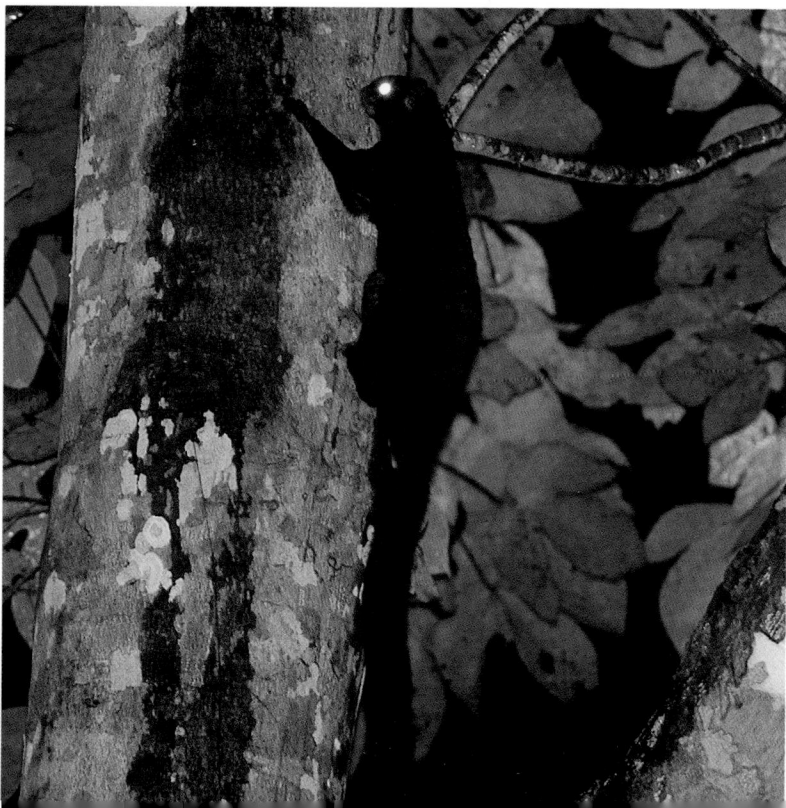

Indian Giant Flying Squirrel ◾ *Petaurista philippensis*
HB 40–49cm, T 40–55cm

DESCRIPTION Upperparts, gliding membrane and tail are grey-brown with long hairs on the back heavily frosted with white. Underparts light grey. Some geographic variation, with upperparts reddish, dark brown or dark grey, with frosting, and underparts pale orange to buff.

DISTRIBUTION Cambodia, Laos, Myanmar, Thailand and Vietnam. Also found in China, India, Sri Lanka and Taiwan.

HABITS AND HABITAT Arboreal and nocturnal species, found up to 1,000m in Southeast Asia. Occurs in hilly and lower montane forests, including primary and tall secondary forests, as well as some orchards and plantation forests.

Spotted Giant Flying Squirrel ■ *Petaurista elegans*

HB 34–36.5cm, T 34–36.5cm

DESCRIPTION Variable. Upperparts, including membrane, dark rufous and black with extensive large white spots, underparts pale rufous and tail black (Peninsular Malaysia and Thailand). Elsewhere in the region the upperparts are brown with fewer white spots, mainly on the head and the centre of the back, and the feet and tail are reddish brown. In Java, white spots may be lacking entirely.

DISTRIBUTION Brunei, Indonesia (Java, Kalimantan, Sumatra), Malaysia (Peninsular Malaysia, Sabah, Sarawak).

HABITS AND HABITAT Nocturnal and strictly arboreal. Found in hill and montane forests, in tall trees and scrub, as well as on rock cliffs. Nests in tree-hollows.

Red Giant Flying Squirrel ■ *Petaurista petaurista*

HB 40–52cm, T 40–60cm

DESCRIPTION Colour and pattern vary across the range of this species. In Peninsular Malaysia the entire body is dark reddish with a clean-cut black tail tip, feet, nose, chin, eye-ring and behind the ears. In southern Myanmar and western Thailand the colour is similar but there is light speckling on the head and back. Variations in other parts of the range include pale orange upperparts with some speckling, whitish underparts, and a reddish, brownish or grey tail.

DISTRIBUTION Brunei, Indonesia (Java, Kalimantan, Sumatra), Malaysia (Peninsular Malaysia, Sabah, Sarawak), Myanmar and Thailand. Also found in China, India and Nepal. Not seen in Singapore since 1995 and may be extirpated.

HABITS AND HABITAT Arboreal and nocturnal, although occasionally active during the day, especially just before dusk. Occurs in a wide variety of forests from lowlands to mountain tops, including in tall secondary forests. Nests in holes in large trees.

Palawan Maxomys ▪ *Maxomys panglima* TL 37–41cm, T 18–23cm

DESCRIPTION Upperparts greyish brown to brown. Underparts white. Pink nose and large ears. The scaly tail is dark above and white below. Very spiny fur – more so than any other rat in the Palawan Faunal Region.

DISTRIBUTION Philippines. Endemic to the Palawan Faunal Region on the islands of Balabac, Palawan, Busuanga, Calauit and Culion.

HABITS AND HABITAT Largely nocturnal. Found in primary and secondary forests, as well as in agricultural areas and plantations. Lowlands and low montane habitats to at least 1,550m.

NOTES This is one of the commonest rats in the Palawan Faunal Region.

Red Spiny Maxomys ▪ *Maxomys surifer* HB 16–21cm, T 15–21cm

DESCRIPTION Upperparts are a rich orange-brown to reddish brown, contrasting with a white underside, with extensive stiff spines amongst the dorsal and ventral fur. The orange-brown colouring extends under the neck, forming a collar. The very long tail is bicoloured, dark above and light below, and is largely naked. The snout is long and pointed.

DISTRIBUTION Brunei, Cambodia, Indonesia (Java, Kalimantan, Sumatra), Laos, Malaysia (Peninsular Malaysia, Sabah, Sarawak), Myanmar, Thailand and Vietnam. Believed to be extirpated in Singapore. Also found in China.

HABITS AND HABITAT Nocturnal and exclusively terrestrial. This species occurs in primary and mature secondary forest, and at forest edges in adjacent cultivated areas. Does not occur in heavily degraded habitats or in large-scale oil-palm plantations.

NOTES There is considerable geographic variation, and more than one species may be represented – more studies are required.

Malaysian Wood Rat ■ *Rattus tiomanicus* HB 14–19cm, T 15–20cm

DESCRIPTION Upperparts finely grizzled olive-brown with short stiff spines and dark guard hairs. Underparts white. Tail dark brown. Large ears. Foot pads are well adapted for climbing.

DISTRIBUTION Brunei, Indonesia (Java, Kalimantan, Sumatra), Malaysia (Peninsular Malaysia, Sabah, Sarawak), Philippines (the Palawan Faunal Region) and Thailand.

HABITS AND HABITAT Found in a variety of lowland habitats, including secondary forest, agricultural areas, plantations, grassland–forest mosaic, selectively logged forests and rice fields.

NOTES This species climbs well and spends considerable time in trees.

Tanezumi Rat ▪ *Rattus tanezumi* HB 10.5–21.5cm, T 12–23cm

DESCRIPTION Upperparts olive-brown to reddish brown, underparts much lighter. The tail is dark grey and mostly naked. The ears are large and the eyes are black. Very similar to the House Rat (*Rattus rattus*) and often lumped together as the same species. However, both these species likely represent a larger species complex.

DISTRIBUTION Cambodia, Laos, Malaysia (Peninsular Malaysia, Sabah, Sarawak), Singapore, Thailand and Vietnam. Introduced in Indonesia and Philippines. Also found in Afghanistan, Bangladesh, Bhutan, China, India, Japan, North and South Korea, Nepal and Taiwan.

HABITS AND HABITAT Nocturnal and sometimes diurnal. Extremely adaptable and commensal with humans, found in many man-made habitats including towns, cities and agricultural areas, feeding on a variety of waste and food scraps, as well as seeds, fruits, insects and some small animals. It is also found in disturbed forests in some areas.

NOTES In some areas, such as the Philippines, this is an extremely destructive introduced pest.

Panay Bushy-tailed Cloud Rat ▪ *Crateromys heaneyi*

HB 28–35cm, T 30–40cm

DESCRIPTION Dark brown to reddish brown, with thick, soft fur. The fur on the underparts is shorter and paler, and the fur on the tail is black. The head is broad with a short snout and a 'mask' of greyish fur on the cheeks, and the eyes and ears are proportionately small.

DISTRIBUTION Philippines. Endemic to the Greater Negros–Panay Faunal Region on Panay Island.

HABITS AND HABITAT Nocturnal and arboreal. Dependent on primary and secondary forests. Found up to 400m, but may occur at higher altitudes. Feeds on fruits and leaves. Nests in holes in large trees.

NOTES Severely impacted by habitat destruction due to illegal logging and agricultural encroachment. There is some hunting pressure. Also known as the Panay Cloud Runner, because of its arboreal habits.

East Asian Porcupine ■ *Hystrix brachyura* HB 59–72cm, T 6–11cm

DESCRIPTION Porcupines, well known for their quills, are globally widespread. The East Asian Porcupine has dark brown to black short hair with long quills on the lower back and shorter quills on the back of the neck and the upper back. Thick, hollow quills on the short tail, which rattle when shaken. Quills are black with pale bases and tips. There is a white band on the upper chest and a blunt face.

DISTRIBUTION Brunei, Cambodia, Indonesia (Kalimantan, Sumatra), Laos, Malaysia (Peninsular Malaysia, Sabah, Sarawak), Myanmar, Singapore, Thailand and Vietnam. Also found in Bangladesh, China, India and Nepal.

HABITS AND HABITAT Terrestrial and largely nocturnal. Found in primary and secondary forests, open areas and cultivated land, but always near rocky, hilly areas, where they dig burrows.

NOTES Burrows are frequently found in rugged outcrops and limestone areas, and are most often occupied by family groups.

Palawan Porcupine ■ *Hystrix pumila* HB 49–60cm, T 4–10cm

DESCRIPTION The body and tail are densely covered with quills. The upperparts are dark brown to black, with whitish underparts. It has a short tail, small ears and eyes.

DISTRIBUTION Philippines. Endemic to the Palawan Faunal Region on the islands of Busuanga and Palawan.

HABITS AND HABITAT It is found in mainly lowland primary and secondary forest. It is nocturnal, and while not much is known of its diet, it is thought to feed on similar items to other members of its genus, such as roots and tubers as well as some animal matter. During the day it shelters in underground burrows, typically shared by members of a small family group.

NOTES When threatened, the Palawan Porcupine may raise its quills, making it appear much larger than it really is. If the threat persists, it may stamp its feet, rattle its tail, and eventually charge backwards, using its quills as a weapon.

▪ PORCUPINES ▪

Long-tailed Porcupine ▪ *Trichys fasciculata* HB 37.5–43.5cm, T 15–24cm

DESCRIPTION Upperparts brown, with short, flattened quills. Underparts pale. Quills are dark but whitish towards the base, and cannot be erected. The tail is long and scaly, with a brush of hollow bristles at the end.

DISTRIBUTION Brunei, Indonesia (Kalimantan, Sumatra) and Malaysia (Peninsular Malaysia, Sabah, Sarawak).

HABITS AND HABITAT Found in lowland and lower montane primary forest, as well as secondary and cultivated areas. Spends time foraging on the ground as well as in trees.

NOTES This is the smallest of the region's porcupines and is somewhat rat-like in appearance.

Taxonomy in this list has been updated to match the Mammal Diversity Database, version 1.12.1 which was released on 30 January 2024 (it can be found from this page: https://www.mammaldiversity.org/).

For each species, an '×' indicates presence in a particular country. A question mark (?) is added if the species is believed to occur in the country but there is some uncertainty. A dagger (†) indicates that the species has been recorded from the country in recent history (within the past 50–100 years), but is now probably (x†) or definitely (†) extirpated from that country.

Country abbreviations are as follows:

PH	Philippines
ID	Indonesia
BN	Brunei
ME	East Malaysia (Sabah, Sarawak)
MP	Peninsular Malaysia
SG	Singapore
MM	Myanmar
TH	Thailand
KH	Cambodia
LA	Laos
VN	Vietnam

In the checklist, marine mammals are listed separately, because the occurrence of marine mammals in the region is poorly understood, and a country-by-country listing was not feasible given their movements and occurrence in international waters. We have listed the marine mammal species that are generally known from Southeast Asian waters.

The known list of species from the region has grown substantially since the first edition of this book, and we can anticipate it will continue to change in the future. New surveys in areas that have been relatively little studied, especially for hard-to-find species such as shrews or bats, regularly yield new distributional records or previously undescribed species. In addition, increased taxonomic work, particularly using tools such as genetics, is resulting in many changes. Often, genetic analyses demonstrate that what was thought to be a single widespread species is actually several different species, although the reverse can sometimes happen. Such genetic studies also lead to new insights into the relationships among species, which can lead to changes in scientific names.

English name	Species name	PH	ID	BN	ME	MP	SG	MM	TH	KH	LA	VN
ARTIODACTYLA												
Bovidae												
Gaur	*Bos gaurus*					x		x	x	x	x	x
Banteng	*Bos javanicus*		x		x	x		x	x	x	x	x
Kouprey	*Bos sauveli*							+	+	x†	x†	+
Wild Water Buffalo	*Bubalus arnee*					+†		x	x	x	+	x†
Tamaraw	*Bubalus mindorensis*	x										
Himalayan Takin	*Budorcas taxicolor*							x				
Burmese Red Serow	*Capricornis rubidus*							x				
Maned Serow	*Capricornis sumatraensis*		x			x		x	x	x	x	x
Cranbrook's Goral	*Naemorhedus cranbrooki*							x				
Chinese Goral	*Naemorhedus griseus*							x				
Myanmar Goral	*Naemorhedus evansi*							x	x			
Blue Sheep	*Pseudois nayaur*							x				
Saola	*Pseudoryx nghetinhensis*										x	x
Cervidae												
Calamian Hog Deer	*Axis calamianensis*	x										
Bawean Hog Deer	*Axis kuhlii*		x									
Indochinese Hog Deer	*Axis porcinus*							x	x	x		x
Sika	*Cervus nippon*											x
Tufted Deer	*Elaphodus cephalophus*							x†				
Bornean Yellow Muntjac	*Muntiacus atherodes*		x	x				x	x			
Fea's Muntjac	*Muntiacus feae*							x	x			
Gongshan Muntjac	*Muntiacus gongshanensis*							x				
Southern Red Muntjac	*Muntiacus muntjak*		x	x	x	x	+	x	x			
Puhoat Muntjac	*Muntiacus puhoatensis*											x
Leaf Muntjac	*Muntiacus putaoensis*							x				
Roosevelt's Muntjac	*Muntiacus rooseveltorum*							x?			x	x?
Annamite Muntjac	*Muntiacus truongsonensis*										x	x
Northern Red Muntjac	*Muntiacus vaginalis*							x	x	x	x	x
Large-antlered Muntjac	*Muntiacus vuquangensis*									x	x	x
Eld's Deer	*Rucervus eldii*							x	x?	x	x	x?
Philippine Spotted Deer	*Rusa alfredi*	x										

English name	Species name	PH	ID	BN	ME	MP	SG	MM	TH	KH	LA	VN
Philippine Brown Deer	Rusa marianna	x										
Javan Rusa	Rusa timorensis		x									
Sambar	Rusa unicolor		x	x	x	x		x	x	x	x	x
Moschidae												
Forest Musk-deer	Moschus berezovskii										?	x
Black Musk-deer	Moschus fuscus							x				x
Suidae												
Palawan Bearded Pig	Sus ahoenobarbus	x										
Bearded Pig	Sus barbatus	x†	x	x	x	x						
Visayan Warty Pig	Sus cebifrons	x										
Oliver's Warty Pig	Sus oliveri	x										
Philippine Warty Pig	Sus philippensis	x										
Eurasian Wild Pig	Sus scrofa		x		x	x		x	x	x	x	x
Javan Warty Pig	Sus verrucosus		x									
Tragulidae												
Javan Chevrotain	Tragulus javanicus		x									
Lesser Oriental Chevrotain	Tragulus kanchil		x	x	x	x	x	x	x	x	x	x
Greater Oriental Chevrotain	Tragulus napu		x	x	x	x	x	x	x			
Balabac Chevrotain	Tragulus nigricans	x										
Silver-backed Chevrotain	Tragulus versicolor											x
Williamson's Chevrotain	Tragulus williamsoni										?	?
CARNIVORA												
Ailuridae												
Eastern Red Panda	Ailurus styani											
Canidae												
Golden Jackal	Canis aureus							x				
Dhole	Cuon alpinus		x			x		x	x	x	x	x‡
Raccoon Dog	Nyctereutes procyonoides											x
Red Fox	Vulpes vulpes							x				x?
Felidae												
Bay Cat	Catopuma badia		x?		x							
Asian Golden Cat	Catopuma temminckii				x			x	x	x	x	x
Jungle Cat	Felis chaus							x	x	x	x	x

English name	Species name	PH	ID	BN	ME	MP	SG	MM	TH	KH	LA	VN
Sunda Clouded Leopard	Neofelis diardi		x	x								x
Mainland Clouded Leopard	Neofelis nebulosa				x			x	x	x	x	x
Leopard	Panthera pardus					x		x	x	x	x	x
Tiger	Panthera tigris		x			x	†	x†	x	x	x	x†
Marbled Cat	Pardofelis marmorata		x		x	x		x	x	x	x	x
Sunda Leopard Cat	Prionailurus javanensis	x	x	x	x							
Mainland Leopard Cat	Prionailurus bengalensis					x	x	x	x	x	x	x
Flat-headed Cat	Prionailurus planiceps		x	x	x	x						
Fishing Cat	Prionailurus viverrinus		x?					x	x	x	x?	x?
Herpestidae												
Short-tailed Mongoose	Urva brachyura		x	x	x	x						
Small Asian Mongoose	Urva auropunctata							x	x?	x	x	x
Javan Mongoose	Urva javanica		x			x		x	x	x	x	x
Collared Mongoose	Urva semitorquatus	x	x	x?	x							
Crab-eating Mongoose	Urva urva		x			x		x	x	x	x	x
Mustelidae												
Greater Hog Badger	Arctonyx albogularis							x				
Northern Hog Badger	Arctonyx collaris							x	x	x	x	x
Sumatran Hog Badger	Arctonyx hoevenii		x									
Oriental Small-clawed Otter	Lutra cinerea	x	x	x	x	x	x	x	x	x	x	x
Eurasian Otter	Lutra lutra		x					x	x	x	x	x
Hairy-nosed Otter	Lutra sumatrana		x	x	x	x		x	x	x	?	x
Smooth Otter	Lutra perspicillata		x	x	x	x	x	x	x	x	x	x
Yellow-throated Marten	Martes flavigula		x	x	x	x		x	x	x	x	x
Stone Marten	Martes foina				x			x				
Bornean Ferret Badger	Melogale everetti				x							
Small-toothed Ferret Badger	Melogale moschata							x	x?		x	x
Javan Ferret Badger	Melogale orientalis		x									
Large-toothed Ferret Badger	Melogale personata							x	x	x	x	x
Cuc Phuong Ferret Badger	Melogale cucphuongensis											x
Yellow-bellied Weasel	Mustela kathiah							x	x	x	x	x
Indonesian Mountain Weasel	Mustela lutreolina		x									
Least Weasel	Mustela nivalis											x

English name	Species name	PH	ID	BN	ME	MP	SG	MM	TH	KH	LA	VN
Malay Weasel	Mustela nudipes		x	x	x	x						
Siberian Weasel	Mustela sibirica							x	x?	x?		
Stripe-backed Weasel	Mustela strigidorsa							x	x		x	x
Mephitidae												
Sunda Stink-badger	Mydaus javanensis		x	x?	x							
Palawan Stink-badger	Mydaus marchei	x										
Ursidae												
Sun Bear	Helarctos malayanus		x	x	x	x	†	x	x	x	x	x
Asian Black Bear	Ursus thibetanus							x	x	x	x	x
Viverridae												
Binturong	Arctictis binturong	x	x	x	x	x	†	x	x	x	x	x
Small-toothed Palm Civet	Arctogalidia trivirgata		x	x	x	x	x	x	x	x	x	x
Owston's Civet	Chrotogale owstoni									x?	x	x
Otter Civet	Cynogale bennettii		x	x	x	x						
Hose's Civet	Diplogale hosei		x	x	x							
Banded Civet	Hemigalus derbyanus		x	x	x	x		x	x	x	x	x
Masked Palm Civet	Paguma larvata		x	x	x	x		x	x	x	x	x
Northern Palm Civet	Paradoxurus hermaphroditus							x	x	x	x	x
Sumatran Palm Civet	Paradoxurus musangus		x			x	x	x	x	x	x	x
Philippine Palm Civet	Paradoxurus philippinensis	x	x	x	x							
Large-spotted Civet	Viverra megaspila					x		x	x	x	x	x
Malay Civet	Viverra tangalunga	x	x	x?	x	x	x					
Large Indian Civet	Viverra zibetha					x	x†	x	x	x	x	x
Small Indian Civet	Viverricula indica		x			x		x	x	x	x	x
Prionodontidae												
Banded Linsang	Prionodon linsang		x		x	x		x	x			
Spotted Linsang	Prionodon pardicolor							x	x	x	x	x
CHIROPTERA												
Craseonycteridae												
Kitti's Hog-nosed Bat	Craseonycteris thonglongyai							x	x			
Emballonuridae												
Small Asian Sheath-tailed Bat	Emballonura alecto	x	x	x	x							
Lesser Sheath-tailed Bat	Emballonura monticola		x	x	x	x		x				

English name	Species name	PH	ID	BN	ME	MP	SG	MM	TH	KH	LA	VN
Pouched Tomb Bat	*Saccolaimus saccolaimus*	x	x	x	x	x	x	x	x	x		x
Long-winged Tomb Bat	*Taphozous longimanus*		x	x	x	x	x	x	x	x	x	
Black-bearded Tomb Bat	*Taphozous melanopogon*	x	x	x	x	x	x		x	x	x	x
Naked-rumped Tomb Bat	*Taphozous nudiventris*							x				
Theobald's Tomb Bat	*Taphozous theobaldi*		x					x	x	x	x	x
Hipposideridae												
South-east Asian Trident Bat	*Aselliscus stoliczkanus*					x		x	x	x	x	x
Dong Bac Trident Bat	*Aselliscus dongbuccanus*											x
Asian Tailless Roundleaf Bat	*Coelops frithii*		x		x	x		x	x	x		x
Philippine Tailless Roundleaf Bat	*Coelops hirsutus*	x										
Malaysian Tailless Roundleaf Bat	*Coelops robinsoni*								x?			
Ha Long Roundleaf Bat	*Hipposideros alongensis*		x		x	x						x
Philippine Dusky Roundleaf Bat	*Hipposideros antricola*	x										
Great Roundleaf Bat	*Hipposideros armiger*					x		x	x	x	x	x
Bornean Dusky Roundleaf Bat	*Hipposideros cf. ater*		x	x	x							
Bicoloured Roundleaf Bat	*Hipposideros bicolor*	x	x	x	x	x	x	x	x			
Short-headed Roundleaf Bat	*Hipposideros breviceps*		x									
Fawn Roundleaf Bat	*Hipposideros cervinus*	x	x	x	x	x						
Ashy Roundleaf Bat	*Hipposideros cineraceus*		x	x		x		x	x	x	x	x
Large Mindanao Roundleaf Bat	*Hipposideros coronatus*	x										
Cox's Roundleaf Bat	*Hipposideros coxi*				x							
Diadem Roundleaf Bat	*Hipposideros diadema*	x	x	x	x	x		x	x	x	x	x
Least Roundleaf Bat	*Hipposideros doriae*		x	x	x	x			x			
Dayak Roundleaf Bat	*Hipposideros dyacorum*		x	x	x	x		x	x			
House-dwelling Roundleaf Bat	*Hipposideros einnaythu*							x	x			
Cantor's Roundleaf Bat	*Hipposideros galeritus*		x	x	x	x		?	x	x	x	x
Andersen's Roundleaf Bat	*Hipposideros gentilis*							x	x	x	x	x
Grand Roundleaf Bat	*Hipposideros grandis*							x	x		x	x
Griffin's Roundleaf Bat	*Hipposideros griffini*											x
Thailand Roundleaf Bat	*Hipposideros halophyllus*							x	x			
Khaokhouay Roundleaf Bat	*Hipposideros khaokhouayensis*										x	
Kingston's Roundleaf Bat	*Hipposideros kingstonae*				x			x	x			
Kunz's Roundleaf Bat	*Hipposideros kunzi*					x		x	x			

145

English name	Species name	PH	ID	BN	ME	MP	SG	MM	TH	KH	LA	VN
Intermediate Roundleaf Bat	Hipposideros larvatus				x	x		x	x	x	x	x
Boonsong's Roundleaf Bat	Hipposideros lekaguli	x							x			
Sheild-faced Roundleaf Bat	Hipposideros byei					x		x	x			x
Maduran Roundleaf Bat	Hipposideros madurae		x									
Malayan Roundleaf Bat	Hipposideros nequam					x						
Philippine Forest Roundleaf Bat	Hipposideros obscurus	x										
Orbiculus Roundleaf Bat	Hipposideros orbiculus		x			x						
Pendlebury's Roundleaf Bat	Hipposideros pendleburyi								x			
Allen's Roundleaf Bat	Hipposideros poutensis							x				x
Philippine Pygmy Roundleaf Bat	Hipposideros pygmaeus	x										
Ridley's Roundleaf Bat	Hipposideros ridleyi			x	x	x			x			
Annamite Roundleaf Bat	Hipposideros rotalis										x	x
Sheild-nosed Roundleaf Bat	Hipposideros scutinares										x	x
Sorensen's Roundleaf Bat	Hipposideros sorenseni		x									
Swinhoe's Roundleaf Bat	Hipposideros swinhoii											x
Megadermatidae												
Thongaree's False-vampire	Eudiscoderma thongareeae								x			
Greater False-vampire	Lyroderma lyra					x		x	x	x	x	x
Lesser False-vampire	Megaderma spasma	x	x	x	x	x	x	x	x	x	x	x
Miniopteridae												
Little Bent-winged Bat	Miniopterus australis	x	x	x	x							
Asian Bent-winged Bat	Miniopterus fuliginosus	x	x	x	x	x		x	x	x	x	x
Large Bent-winged Bat	Miniopterus magnater		x		x	x		x	x	x	x	x
Medium Bent-winged Bat	Miniopterus medius		x	x	x	x			x			
Philippine Bent-winged Bat	Miniopterus paululus	x	x		x							
Small Bent-winged Bat	Miniopterus pusillus		x					x	x	x	x	x
Great Bent-winged Bat	Miniopterus tristis	x										
Molossidae												
Lesser Naked Bat	Cheiromeles parvidens	x										
Greater Naked Bat	Cheiromeles torquatus	x	x	x	x	x	x		x			
Johore Wrinkle-lipped Bat	Mops johorensis		x			x			x			
Sunda Free-tailed Bat	Mops mops		x	x	x	x			x			
Asian Wrinkle-lipped Bat	Mops plicata	x	x	x	x	x		x	x	x	x	x

English name	Species name	PH	ID	BN	ME	MP	SG	MM	TH	KH	LA	VN
Sulawesi Free-tailed Bat	Mops sarasinorum	x	x									
Sumatran Mastiff Bat	Mormopterus doriae		x									
Java Giant Mastiff Bat	Otomops formosus		x									
Wroughton's Giant Mastiff Bat	Otomops wroughtoni									x		
La Touche's Free-tailed Bat	Tadarida latouchei								x		x	x
Nycteridae												
Javan Slit-faced Bat	Nycteris javanica		x									
Malayan Slit-faced Bat	Nycteris tragata		x	x	x	x	x	x	x			
Pteropodidae												
Golden-crowned Flying-fox	Acerodon jubatus	x										
Palawan Flying-fox	Acerodon leucotis	x										
Bornean Hill Fruit Bat	Aethalops aequalis		x	x	x							
Grey Fruit Bat	Aethalops alecto		x			x						
Mindanao Fruit Bat	Alionycteris paucidentata	x										
Bornean Spotted-winged Fruit Bat	Balionycteris maculata		x	x	x							
Malayan Spotted-winged Fruit Bat	Balionycteris seimundi					x						
Sundaic Black-capped Fruit Bat	Chironax melanocephalus		x	x	x	x						
Lesser Short-nosed Fruit Bat	Cynopterus brachyotis	x?	x	x	x	x	x	x	x	x	x	x
Horsfield's Fruit Bat	Cynopterus horsfieldii		x	x	x	x			x	x	x	x
Peters's Fruit Bat	Cynopterus luzoniensis	x										
Least Fruit Bat	Cynopterus minutus		x	x	x	x						
Greater Short-nosed Fruit Bat	Cynopterus sphinx		x			x		x	x	x	x	x
Indonesian Short-nosed Fruit Bat	Cynopterus titthaecheilus		x									
Negros Naked-backed Fruit Bat	Dobsonia chapmani	x										
Brooks's Fruit Bat	Dyacopterus brooksi		x									
Dayak Fruit Bat	Dyacopterus spadiceus	x	x	x	x	x			x			
Philippine Large-headed Fruit Bat	Dyacopterus rickarti	x										
Greater Nectar Bat	Eonycteris major		x	x	x							
Philippine Nectar Bat	Eonycteris robusta	x										
Cave Nectar Bat	Eonycteris spelaea	x	x	x	x	x	x	x	x	x	x	x
Philippine Pygmy Fruit Bat	Haplonycteris fischeri	x										
Philippine Harpy Fruit Bat	Harpyionycteris whiteheadi	x										
Lesser Long-tongued Nectar Bat	Macroglossus minimus	x	x	x	x	x	x	x	x	x	x	x

English name	Species name	PH	ID	BN	ME	MP	SG	MM	TH	KH	LA	VN
Greater Long-tongued Nectar Bat	Macroglossus sobrinus		x			x		x	x	x	x	x
Sunda Tailless Fruit Bat	Megaerops ecaudatus		x	x	x	x			x			
Javan Tailless Fruit Bat	Megaerops kusnotoi		x									
Northern Tailless Fruit Bat	Megaerops niphanae							x	x	x	x	x
White-collared Fruit Bat	Megaerops albicollis		x	x	x	x						
Philippine Tube-nosed Fruit Bat	Nyctimene rabori	x										
Luzon Fruit Bat	Otopteropus cartilagonodus	x										
Dusky Fruit Bat	Penthetor lucasi		x	x	x	x	x	x	x			
Greater Musky Fruit Bat	Ptenochirus jagori	x										
Lesser Musky Fruit Bat	Ptenochirus minor	x										
Mindanao Musky Fruit Bat	Ptenochirus wetmorei	x										
Ryukyu Flying-fox	Pteropus dasymallus	x										
Anderson's Flying-fox	Pteropus intermedius							x				
Island Flying-fox	Pteropus hypomelanus	x	x					x	x	x		x
Lyle's Flying-fox	Pteropus lylei								x	x		x
Black-eared Flying-fox	Pteropus melanotus	x	x									
Little Golden-mantled Flying-fox	Pteropus pumilus	x										
Philippine Grey Flying-fox	Pteropus speciosus	x	x									
Large Flying-fox	Pteropus vampyrus	x	x	x	x	x	x	x	x	x		x
Dwarf Flying-fox	Pteropus woodfordi	x										
Geoffroy's Rousette	Rousettus amplexicaudatus	x	x	x	x	x	x	x	x	x	x	x
Leschenault's Rousette	Rousettus leschenaultii					x		x	x	x	x	x
Bare-backed Rousette	Rousettus spinalatus		x	x	x							
Hill Fruit Bat	Sphaerias blanfordi							x	x			
Wallace's Stripe-faced Fruit Bat	Styloctenium mindorensis	x										
Rhinolophidae												
Acuminate Horseshoe Bat	Rhinolophus acuminatus		x	x	x	x	x	x	x	x	x	x
Intermediate Horseshoe Bat	Rhinolophus affinis		x		x	x		x	x	x	x	x
Arcuate Horseshoe Bat	Rhinolophus arcuatus	x	x									
Bornean Horseshoe Bat	Rhinolophus borneensis		x	x	x							
Canut's Horseshoe Bat	Rhinolophus canuti		x									
Sulawesi Horseshoe Bat	Rhinolophus celebensis		x									
Indochinese Horseshoe Bat	Rhinolophus chaseni									x	x	x

English name	Species name	PH	ID	BN	ME	MP	SG	MM	TH	KH	LA	VN
Chiew Kwee's Horseshoe Bat	Rhinolophus chiewkweeae					x						
Chutamas's Horseshoe Bat	Rhinolophus chutamasae								x		x	
Croslet Horseshoe Bat	Rhinolophus coelophyllus					x		x	x		x	
Convex Horseshoe Bat	Rhinolophus convexus					x					x	
Creagh's Horseshoe Bat	Rhinolophus creaghi	x	x	x	x							
Allen's Horseshoe Bat	Rhinolophus episcopus											x
Andersen's Woolly Horseshoe Bat	Rhinolophus foetidus		x	x		x						
Francis's Woolly Horseshoe Bat	Rhinolophus francisi		x		x				x			x
Philippine Big-eared Horseshoe Bat	Rhinolophus hirsutus	x										
Philippine Forest Horseshoe Bat	Rhinolophus inops	x										
Blyth's Horseshoe Bat	Rhinolophus lepidus		x					x	x	x		x
Selangor Woolly Horseshoe Bat	Rhinolophus luctoides					x						
Great Woolly Horseshoe Bat	Rhinolophus luctus		x									
Big-eared Horseshoe Bat	Rhinolophus macrotis		x			x		x	x		x	x
Madura Hoseshoe Bat	Rhinolophus madurensis		x									
Malayan Horseshoe Bat	Rhinolophus malayanus					x		x	x	x	x	x
Marshall's Horseshoe Bat	Rhinolophus marshalli					x		x	x	x	x	x
Indochinese Brown Horseshoe Bat	Rhinolophus microglobosus					x		x	x	x	x	x
Malaysian Woolly Horseshoe Bat	Rhinolophus morio					x		x	x			
Osgood's Horseshoe Bat	Rhinolophus osgoodi											x
Pearson's Horseshoe Bat	Rhinolophus pearsonii							x	x	x	x	x
Northern Woolly Horseshoe Bat	Rhinolophus perniger							x	x	x	x	x
Large-eared Horseshoe Bat	Rhinolophus philippinensis	x	x	x	x							
Sarawak Horseshoe Bat	Rhinolophus proconsulis				x							
Least Horseshoe Bat	Rhinolophus pusillus		x			x		x	x	x	x	x
Glossy Horseshoe Bat	Rhinolophus refulgens		x			x	x		x			
King Horseshoe Bat	Rhinolophus rex								x			
Peninsular Horseshoe Bat	Rhinolophus robinsoni					x			x		x	x
Indian Rufous Horseshoe Bat	Rhinolophus rouxii							x				
Large Rufous Horseshoe Bat	Rhinolophus rufus	x										
Lesser Woolly Horseshoe Bat	Rhinolophus sedulus		x	x	x							
Shamel's Horseshoe Bat	Rhinolophus shameli								x	x	x	x
Shortridge's Horseshoe Bat	Rhinolophus shortridgei							x				

English name	Species name	PH	ID	BN	ME	MP	SG	MM	TH	KH	LA	VN
Siam Horseshoe Bat	Rhinolophus siamensis								x		x	x
Chinese Horseshoe Bat	Rhinolophus sinicus											x
Lesser Brown Horseshoe Bat	Rhinolophus stheno					x	x	x	x		x	x
Little Nepalese Horseshoe Bat	Rhinolophus subbadius		x					x				
Small Rufous Horseshoe Bat	Rhinolophus subrufus	x										
Thailand Horseshoe Bat	Rhinolophus thailandensis							x	x			
Thomas's Horseshoe Bat	Rhinolophus thomasi							x	x		x	x
Trefoil Horseshoe Bat	Rhinolophus trifoliatus		x	x	x	x	x	x	x			
Yellow-faced Horseshoe Bat	Rhinolophus virgo	x										
Yong Hoi Sen's Horseshoe Bat	Rhinolophus yonghoiseni					x						
Dobson's Horseshoe Bat	Rhinolophus yunanensis							x				
Rhinopomatidae												
Lesser Mouse-tailed Bat	Rhinopoma hardwickii								x			
Greater Mouse-tailed Bat	Rhinopoma microphyllum		x					x?	x?			
Vespertilionidae (Kerivoulinae)												
Flores Woolly Bat	Kerivoula flora	x	x		x							
Flat-skulled Woolly Bat	Kerivoula depressa							x	x	x	x	x
Dark Woolly Bat	Kerivoula furva							x				x
Hardwicke's Woolly Bat	Kerivoula hardwickii	x	x	x	x	x	x	x	x	x	x	x
Small Woolly Bat	Kerivoula intermedia		x	x	x	x						
Kachin Woolly Bat	Kerivoula kachinensis							x	x	x	x	x
Krau Woolly Bat	Kerivoula krauensis			x		x						
Indian Woolly Bat	Kerivoula lenis		x	x	x	x						
Least Woolly Bat	Kerivoula minuta		x	x	x	x			x			
Papillose Woolly Bat	Kerivoula papillosa	x	x	x	x	x			x	x	x	x
Clear-winged Woolly Bat	Kerivoula pellucida	x	x	x	x	x			x			
Painted Woolly Bat	Kerivoula picta		x			x		x	x	x	x	x
Bornean Woolly Bat	Kerivoula pusilla		x	x	x				x			
Titania's Woolly Bat	Kerivoula titania							x	x	x	x	x
Whitehead's Woolly Bat	Kerivoula whiteheadi	x										
Lesser Groove-toothed Bat	Phoniscus atrox		x	x	x	x			x			
Greater Groove-toothed Bat	Phoniscus jagorii	x	x	x	x	x		x	x		x	x

English name	Species name	PH	ID	BN	ME	MP	SG	MM	TH	KH	LA	VN
Vespertilionidae (Murininae)												
Hairy-winged Bat	Harpiocephalus harpia	x			x	x		x	x	x	x	x
Formosan Tube-nosed Bat	Harpiola isodon				x	x						x
Bronzed Tube-nosed Bat	Murina aenea		x	x	x			x	x			
Annamite Tube-nosed Bat	Murina annamitica								x	x	x	x
Bala Tube-nosed Bat	Murina balaensis					x			x			
Beelzebub Tube-nosed Bat	Murina beelzebub											x
Golden-haired Tube-nosed Bat	Murina chrysochaetes											x
Round-eared Tube-nosed Bat	Murina cyclotis	x						x	x	x	x	x
Elery's Tube-nosed Bat	Murina eleryi							x	x		x	x
Fea's Tube-nosed Bat	Murina feae							x	x	x	x	x
Fiona's Tube-nosed Bat	Murina fionae									x		x
Guillen's Tube-nosed Bat	Murina guilleni								x			
Vietnamese Tube-nosed Bat	Murina harpioloides											x
Harrison's Tube-nosed Bat	Murina harrisoni								x	x	x	x
Hkakabo Razi Tube-nosed Bat	Murina hkakaboraziensis							x				
Hutton's Tube-nosed Bat	Murina huttoni					x		x	x		x	x
Jiantia Tube-nosed Bat	Murina jiantiana							x				
Kon Tum Tube-nosed Bat	Murina kontumensis											x
Greater Tube-nosed Bat	Murina leucogaster											x
Lorelie's Tube-nosed Bat	Murina lorelieae											x
Peninsular Tube-nosed Bat	Murina peninsularis		x	x	x	x			x			
Rozendaal's Tube-nosed Bat	Murina rozendaali		x	x	x	x			x			
Lesser Tube-nosed Bat	Murina suilla	x	x	x	x	x	x		x			
Walston's Tube-nosed Bat	Murina walstoni								x	x	x	x
Vespertilionidae (Myotinae)												
Disc-footed Bat	Eudiscopus denticulatus							x	x		x	x
Grey Large-footed Myotis	Myotis adversus		x	x	x	x	x					
Szechuan Myotis	Myotis altarium								x			x
Indochinese Whiskered Myotis	Myotis alticraniatus							x	x	x	x	x
Valley Myotis	Myotis ancricola							x			x	x
Hairy-faced Myotis	Myotis annectans								x	x	x	x
Peters's Myotis	Myotis ater	x							x	x	x	x

English name	Species name	PH	ID	BN	ME	MP	SG	MM	TH	KH	LA	VN
Bartel's Myotis	Myotis bartelsii		x									
Bornean Whiskered Myotis	Myotis borneoensis		x									x
Chinese Myotis	Myotis chinensis				x							
Malayan Whiskered Myotis	Myotis federatus					x		x	x		x	
Black-and-orange Myotis	Myotis formosus	x	x								x	
Gomantong Myotis	Myotis gomantongensis		x		x							
Hasselt's Myotis	Myotis hasseltii		x	x	x	x		x	x	x		x
Hayes's Thick-thumbed Myotis	Myotis hayesi									x		
Herman's Myotis	Myotis hermani		x			x			x			
Horsefield's Myotis	Myotis horsefieldii	x	x	x	x	x	x	x	x	x	x	x
Indochinese Myotis	Myotis indochinensis	x	x			x					x	x
Chinese Water Myotis	Myotis laniger										x	x
Pallid Large-footed Myotis	Myotis macrotarsus	x	x		x							
Burmese Whiskered Myotis	Myotis montivagus							x				
Asian Whiskered Myotis	Myotis muricola	x	x	x	x	x	x	x	x	x	x	x
Nepal Myotis	Myotis nipalensis							x				
Phan Luong's Myotis	Myotis phanluongi											x
Rickett's Myotis	Myotis pilosus										x	x
Ridley's Myotis	Myotis ridleyi	x	x	x	x	x			x			
Thick-thumbed Myotis	Myotis rosseti		x						x	x	x	x
Reddish-black Myotis	Myotis rufoniger										x	x
Orange-fingered Myotis	Myotis rufopictus	x										
Small-toothed Myotis	Myotis siligorensis							x				
Vesptilionidae (Vespertilioninae)												
Collared Sprite	Arielulus aureocollaris								x		x	x
Black Gilded Sprite	Arielulus circumdatus		x			x		x	x	x	x	x
Coppery Sprite	Arielulus cuprosus			x								
Benom Sprite	Arielulus societatis					x						
Eastern Barbastelle	Barbastella darjelingensis							x			x	x
Surat Helmeted Bat	Cassistrellus dimissus								x		x	
Yok Don Helmeted Bat	Cassistrellus yokdonensis										x	
Oriental Serotine	Cnephaeus pachyomus							x	x		x	x
Thick-eared Serotine	Cnephaeus pachyotis							x	x			x

English name	Species name	PH	ID	BN	ME	MP	SG	MM	TH	KH	LA	VN
Dark Thick-thumbed Pipistrelle	Glischropus aquilus		x									
Indochinese Thick-thumbed Pipistrelle	Glischropus bucephalus							x	x	x	x	x
Javan Thick-thumbed Pipistrelle	Glischropus javanus		x									
Sunda Thick-thumbed Pipistrelle	Glischropus tylopus	x	x	x					x			
Least False-serotine	Hesperoptemus blanfordi		x		x	x		x	x	x	x	x
Doria's False-serotine	Hesperoptemus doriae				x	x						
Tickell's False-serotine	Hesperoptemus tickelli							x	x	x	x	x
Tomes's False-serotine	Hesperoptemus tomesi				x	x		x	x			
Chocolate Pipistrelle	Hypsugo affinis							x				
Cadorna's Pipistrelle	Hypsugo cadornae							x	x	x	x	x
Long-toothed Pipistrelle	Hypsugo dolichodon							x	x			x
Imbricate Pipistrelle	Hypsugo imbricatus		x		x	x						
Red-brown Pipistrelle	Hypsugo kitcheneri		x									
Myanmar Pipistrelle	Hypsugo lophurus							x				
Big-eared Pipistrelle	Hypsugo macrotis		x			x	x					
Pungent Pipistrelle	Hypsugo mordax		x									
Peters's Pipistrelle	Hypsugo petersi	x	x		x							
Chinese Pipistrelle	Hypsugo pulveratus							x	x		x	x
White-winged Pipistrelle	Hypsugo vordermanni		x	x	x			x	x		x	x
Great Evening Bat	Ia io							x				
Joffre's Pipistrelle	Mirostrellus joffrei							x				
Chinese Noctule	Nyctalus labiatus							x				x
Narrow-winged Brown Bat	Philetor brachypterus	x	x	x	x	x						
Japanese Pipistrelle	Pipistrellus abramus							x			x	x
Kelaart's Pipistrelle	Pipistrellus ceylonicus							x			x	x
Coromandel Pipistrelle	Pipistrellus coromandra				x			x	x	x	x	x
Javan Pipistrelle	Pipistrellus javanicus	x	x	x	x	x	x	x	x	x	x	x
Mount Popa Pipistrelle	Pipistrellus paterculus					x		x	x	x	x	x
Narrow-winged Pipistrelle	Pipistrellus stenopterus		x		x	x	x	x	x		x	x
Least Pipistrelle	Pipistrellus tenuis	x	x	x				x	x	x	x	x
Harlequin Bat	Scotomanes ornatus	x	x	x	x	x		x	x	x	x	x
Greater Asian House Bat	Scotophilus heathii							x	x	x	x	x
Lesser Asian House Bat	Scotophilus kuhlii	x	x	x	x	x	x	x	x	x	x	x

English name	Species name	PH	ID	BN	ME	MP	SG	MM	TH	KH	LA	VN
Indomalayan Lesser Bamboo Bat	*Tylonycteris fulvida*					x	x	x	x	x	x	x
Malayan Greater Bamboo Bat	*Tylonycteris malayana*					x	x	x	x	x	x	x
Sunda Lesser Bamboo Bat	*Tylonycteris pachypus*	x	x	x	x							
Sunda Greater Bamboo Bat	*Tylonycteris robustula*	x	x	x	x							
Tonkin Greater Bamboo Bat	*Tylonycteris tonkinensis*										x	x
DERMOPTERA												
Cynocephalidae												
Philippine Colugo	*Cynocephalus volans*	x										
Sunda Colugo	*Galeopterus variegatus*		x	x	x	x	x	x	x	x	x	x
INSECTIVORA												
Erinaceidae												
Moonrat	*Echinosorex gymnura*		x	x	x	x	x	x	x			
Bornean Short-tailed Gymnure	*Hylomys dorsalis*		x	x	x							
Dalat Gymnure	*Hylomys macarong*											x
Max's Short-tailed Gymnure	*Hylomys maxi*		x			x						
Dwarf Gymnure	*Hylomys parvus*		x									
Northern Short-tailed Gymnure	*Hylomys peguensis*					x		x	x	x	x	x
Javan Short-tailed Gymnure	*Hylomys suillus*		x									
Leuser Gymnure	*Hylomys vorax*		x									
Chinese Gymnure	*Neotetracus sinensis*							x				x
Large-eared Gymnure	*Otohylomys megalotis*										x	
Dinagat Gymnure	*Podogymnura aureospinula*	x										
Eastern Mindanao Gymnure	*Podogymnura intermedia*	x										
Kitanglad Gymnure	*Podogymnura minima*	x										
Mindanao Gymnure	*Podogymnura truei*	x										
Soricidae												
Assam Mole Shrew	*Anourosorex assamensis*							x				
Chinese Mole Shrew	*Anourosorex squamipes*							x	x		x	x
Myanmar Short-tailed Shrew	*Blarinella wardi*							x				
Malayan Water Shrew	*Chimarrogale hantu*					x						
Himalayan Water Shrew	*Chimarrogale himalayica*							x			x	x
Bornean Water Shrew	*Chimarrogale phaeura*		x?	x								
Styan's Water Shrew	*Chimarrogale styani*							x				

English name	Species name	PH	ID	BN	ME	MP	SG	MM	TH	KH	LA	VN
Sumatran Water Shrew	Chimarrogale sumatrana		x									
Vietnamese Water Shrew	Chimarrogale varennei											x
Van Sung's Brown-toothed Shrew	Chodsigoa caovansunga											x
Dusky Brown-toothed Shrew	Chodsigoa furva							x				
Hoffmann's Brown-toothed Shrew	Chodsigoa hoffmanni											x
Lowe's Shrew	Chodsigoa parca							x	x			x
Javan Long-tailed Shrew	Crocidura absconduta		x									
Annamite Shrew	Crocidura annamitensis											x
Asian Grey Shrew	Crocidura attenuata							x	x		x	x
Kinabalu Shrew	Crocidura baluensis				x							
Batak Shrew	Crocidura batakorum	x										
Lesser Mindanao Shrew	Crocidura beata	x										
Beccari's Shrew	Crocidura beccarii		x									
Thick-tailed Shrew	Crocidura brunnea		x									
Cranbrook's Shrew	Crocidura cranbrooki							x				
Large Shrew	Crocidura dracula							x			x	x
Bornean Shrew	Crocidura foetida			x	x							
South-east Asian Shrew	Crocidura fuliginosa		x			x	x	x	x	x	x	x
Greater Mindanao Shrew	Crocidura grandis	x										
Luzon Shrew	Crocidura grayi	x										
Vietnamese Shrew	Crocidura guy											x
Hill's Shrew	Crocidura hilliana								x		x	
Hutan Shrew	Crocidura hutanis		x									
Indochinese Shrew	Crocidura indochinensis							x	x		x	x
Ke Go Shrew	Crocidura kegoensis											x
Sumatran Giant Shrew	Crocidura lepidura		x									
Malayan Shrew	Crocidura malayana					x						
Javanese Shrew	Crocidura maxi		x									
Mindoro Shrew	Crocidura mindorus	x										
Sunda Shrew	Crocidura monticola		x			x?						
Neglected Shrew	Crocidura neglecta		x									
Peninsular Shrew	Crocidura negligens							x	x			
Negros Shrew	Crocidura negrina	x										

English name	Species name	PH	ID	BN	ME	MP	SG	MM	TH	KH	LA	VN
Sibuyan Shrew	Crocidura ninoyi											.
Oriental Shrew	Crocidura orientalis		x									
Palawan Shrew	Crocidura palawanensis	x										
Panay Shrew	Crocidura panayensis	x										
Sumatran Long-tailed Shrew	Crocidura paradoxura		x									
Phan Luong Shrew	Crocidura phanluongi											x
Phu Quoc Shrew	Crocidura phuquocensis									x		x
Chinese Shrew	Crocidura rapax											x
Sa Pa Shrew	Crocidura sapaensis											x
Sokolov Shrew	Crocidura sokolovi											x
Taiwanese Grey Shrew	Crocidura tanakae										x	x
Javan Ghost Shrew	Crocidura umbra		x									
Voracious Shrew	Crocidura vorax								x		x	x
Banka Shrew	Crocidura vosmaeri		x									
Wuchih Shrew	Crocidura wuchihensis											x
Bailey's Brown-toothed Shrew	Episoriculus baileyi							x				x
Long-tailed Brown-toothed Shrew	Episoriculus macrurus							x				x
Hidden Brown-toothed Shrew	Episoriculus umbrinus							x				
Sikkim Water Shrew	Nectogale sikkimensis							x				
Black Shrew	Palawanosorex ater				x							
Palawan Moss Shrew	Palawanosorex muscorum	x										
Indochinese Short-tailed Shrew	Parablarinella griselda							x?			?	x
Ward's Stripe-backed Shrew	Sorex wardi							x				
Lesser Large-clawed Shrew	Soriculus minor							x				
Pygmy White-toothed Shrew	Suncus etruscus					x		x	x		x	x
Hose's Shrew	Suncus hosei		?	x	x	x						
Malayan Pygmy Shrew	Suncus malayanus					x			x			
House Shrew	Suncus murinus		x	x	x	x	x	x	x	x	x	x
Talpidae												
Large Chinese Mole	Euroscaptor grandis							x				
Kloss's Mole	Euroscaptor klossi								x		x	
Kuznetsov's Mole	Euroscaptor kuznetsovi											
Malaysian Mole	Euroscaptor malayanus					x						x

English name	Species name	PH	ID	BN	ME	MP	SG	MM	TH	KH	LA	VN
Ngoc Linh Mole	Euroscaptor ngoclinhensis											x
Orlov's Mole	Euroscaptor orlovi											x
Small-toothed Mole	Euroscaptor parvidens											x
Vietnamese Mole	Euroscaptor subanura											x
La Touche's Mole	Mogera latouchei											x
Blyth's Mole	Parascaptor leucurus							x				
Long-tailed Mole	Scaptonyx fusicauda							x				x
Fansipan Shrew Mole	Uropsilus fansipanensis											x
Inquisitive Shrew Mole	Uropsilus investigator							x?				
Snow Mountain Shrew Mole	Uropsilus nivatus							x?				
LAGOMORPHA												
Leporidae												
Yunnan Hare	Lepus comus							x				
Burmese Hare	Lepus peguensis							x	x	x	x	x
Chinese Hare	Lepus sinensis											x
Sumatran Striped Rabbit	Nesolagus netscheri		x									
Annamite Striped Rabbit	Nesolagus timminsi										x	x
Ochotonidae												
Forrest's Pika	Ochotona forresti							x				
Mountain Pika	Ochotona thibetana							x				
PERISSODACTYLA												
Rhinocerotidae												
Sumatran Rhinoceros	Dicerorhinus sumatrensis		x		†	†		†				
Javan Rhinoceros	Rhinoceros sondaicus		x									†
Tapiridae												
Asian Tapir	Tapirus indicus		x			x		x	x			
PHOLIDOTA												
Manidae												
Philippine Pangolin	Manis culionensis	x										
Sunda Pangolin	Manis javanica		x	x	x	x	x	x	x	x	x	x
Chinese Pangolin	Manis pentadactyla							x	x		x	x

English name	Species name	PH	ID	BN	ME	MP	SG	MM	TH	KH	LA	VN
PRIMATES												
Tarsiidae												
Philippine Tarsier	*Carlito syrichta*	x										
Western Tarsier	*Cephalopachus bancanus*		x	x	x							
Lorisidae												
Bengal Slow Loris	*Nycticebus bengalensis*							x	x	x	x	x
Sunda Slow Loris	*Nycticebus coucang*		x			x	x		x			
Hiller's Slow Loris	*Nycticebus hilleri*		x									
Javan Slow Loris	*Nycticebus javanicus*		x									
Philippine Slow Loris	*Nycticebus menagensis*	x	x	x	x							
Bangka Slow Loris	*Nycticebus bancanus*		x									
Bornean Slow Loris	*Nycticebus borneanus*		x									
Kayan Slow Loris	*Nycticebus kayan*		x	x	x							
Northern Pygmy Slow Loris	*Xanthonycticebus intermedius*										x	x
Southern Pygmy Slow Loris	*Xanthonycticebus pygmaeus*									x	x	x
Cercopithecidae (Colobines)												
Proboscis Monkey	*Nasalis larvatus*		x	x	x							
Black-and-White Langur	*Presbytis bicolor*		x									
Miller's Grizzled Langur	*Presbytis canicrus*		x									
Bornean Banded Langur	*Presbytis chrysomelas*		x	x	x							
Javan Grizzled Langur	*Presbytis comata*		x									
Common Banded Langur	*Presbytis femoralis*		x			x	x					
White-fronted Langur	*Presbytis frontata*		x			x						
Hose's Langur	*Presbytis hosei*		x	x	x							
Black-crested Sumatran Langur	*Presbytis melalophos*		x									
Mitred Langur	*Presbytis mitrata*		x									
Natuna Islands Langur	*Presbytis natunae*		x									
East Sumatran Banded Langur	*Presbytis percura*		x									
Pagai Langur	*Presbytis potenziani*		x									
Robinson's Banded Langur	*Presbytis robinsoni*					x		x	x			
Maroon Langur	*Presbytis rubicunda*		x	x				x				
Sabah Grizzled Langur	*Presbytis sabana*							x				
White-thighed Langur	*Presbytis siamensis*		x			x			x			

English name	Species name	PH	ID	BN	ME	MP	SG	MM	TH	KH	LA	VN
Siberut Langur	Presbytis siberu		x									
Black Sumatran Langur	Presbytis sumatrana		x									
Thomas's Langur	Presbytis thomasi		x									
Grey-shanked Douc	Pygathrix cinerea									x?		x
Red-shanked Douc	Pygathrix nemaeus									x	x	x
Black-shanked Douc	Pygathrix nigripes									x		x
Tonkin Snub-nosed Monkey	Rhinopithecus avunculus											x
Myanmar Snub-nosed Monkey	Rhinopithecus strykeri							x				
Pig-tailed Langur	Simias concolor		x									
Eastern Ebony Langur	Trachypithecus auratus		x									
Tenasserim Langur	Trachypithecus barbei							x	x			
Indochinese Grey Langur	Trachypithecus crepusculus					x		x	x		x	x
Sundaic Silvered Langur	Trachypithecus cristatus		x	x	x	x						
Delacour's Langur	Trachypithecus delacouri											x
Black Langur	Trachypithecus ebenus										x	x
Francois's Langur	Trachypithecus francoisi											x
Indochinese Silvered Langur	Trachypithecus germaini					x			x	x	x	x
Hatinh Langur	Trachypithecus hatinhensis										x	x
Lao Langur	Trachypithecus laotum										x	
Annamese Langur	Trachypithecus margarita									x	x	x
Ebony Langur	Trachypithecus mauritius		x									
Shan Langur	Trachypithecus melamera							x	x			
Dusky Langur	Trachypithecus obscurus					x		x	x			
Phayre's Langur	Trachypithecus phayrei							x				
Capped Langur	Trachypithecus pileatus							x				
Cat Ba Langur	Trachypithecus poliocephalus											x
Popa Langur	Trachypithecus popa							x				
Selangor Silvered Langur	Trachypithecus selangorensis					x						
Shortridge's Langur	Trachypithecus shortridgei							x				
Cercopithecidae (Macaques)												
Stump-tailed Macaque	Macaca arctoides					x		x	x	x	x	x
Assamese Macaque	Macaca assamensis							x	x		x	x
Long-tailed Macaque	Macaca fascicularis	x	x	x	x	x	x	x	x	x	x	x

English name	Species name	PH	ID	BN	ME	MP	SG	MM	TH	KH	LA	VN
Northern Pig-tailed Macaque	*Macaca leonina*							x	x	x	x	x
Rhesus Macaque	*Macaca mulatta*							x	x		x	x
Southern Pig-tailed Macaque	*Macaca nemestrina*		x	x	x			x	x			
Pagai Island Macaque	*Macaca pagensis*		x									
Siberut Macaque	*Macaca siberu*		x									
Hylobatidae												
Western Hoolock Gibbon	*Hoolock hoolock*							x				
Eastern Hoolock Gibbon	*Hoolock leuconedys*							x				
Skywalker Hoolock Gibbon	*Hoolock tianxing*							x				
Abbott's Grey Gibbon	*Hylobates abbotti*		x		x							
Agile Gibbon	*Hylobates agilis*		x			x			x			
Bornean White-bearded Gibbon	*Hylobates albibarbis*		x									
East Bornean Grey Gibbon	*Hylobates funereus*		x	x	x							
Kloss's Gibbon	*Hylobates klossii*		x									
White-handed Gibbon	*Hylobates lar*		x			x		x	x			
Javan Gibbon	*Hylobates moloch*		x									
Mueller's Bornean Gibbon	*Hylobates muelleri*		x									
Pileated Gibbon	*Hylobates pileatus*								x	x	x	
Northern Yellow-cheeked Gibbon	*Nomascus annamensis*									x	x	x
Western Black Crested Gibbon	*Nomascus concolor*										x	x
Southern Yellow-cheeked Gibbon	*Nomascus gabriellae*									x		x
Northern White-cheeked Gibbon	*Nomascus leucogenys*											x
Eastern Crested Gibbon	*Nomascus nasutus*											x
Southern White-cheeked Gibbon	*Nomascus siki*										x	x
Siamang	*Symphalangus syndactylus*		x			x			x			
Hominidae												
Bornean Orangutan	*Pongo pygmaeus*		x		x							
Sumatran Orangutan	*Pongo abelii*		x									
Tapanuli Orangutan	*Pongo tapanuli*		x									
PROBOSCIDEA												
Elephantidae												
Asian Elephant	*Elephas maximus*		x		x	x		x	x	x	x	x

English name	Species name	PH	ID	BN	ME	MP	SG	MM	TH	KH	LA	VN
RODENTIA												
Cricetidae												
Yunnan Red-backed Vole	Eothenomys eleusis							x	x		x	x
Clarke's Vole	Neodon clarkei							x				x
Forrest's Mountain Vole	Neodon forresti							x				
Diatomyidae												
Kha-nyou	Laonastes aenigmamus										x	
Hystricidae												
Asian Brush-tailed Porcupine	Atherurus macrourus		x	x		x		x	x	x	x	x
East Asian Porcupine	Hystrix brachyura		x	x	x	x	x	x	x	x	x	x
Thick-spined Porcupine	Hystrix crassispinis		x	x	x							
Sunda Porcupine	Hystrix javanica		x									
Palawan Porcupine	Hystrix pumila	x										
Sumatran Porcupine	Hystrix sumatrae		x									
Long-tailed Porcupine	Trichys fasciculata		x	x	x							
Muridae												
Luzon Broad-toothed Rat	Abditomys latidens	x										
Mindoro Climbing Rat	Anonymomys mindorensis	x										
Striped Field Mouse	Apodemus agrarius							x				
Yunnan Wood Mouse	Apodemus ilex							x				
Large-eared Wood Mouse	Apodemus latronum							x				
Luzon Cordillera Forest Mouse	Apomys abrae	x										
Aurora Forest Mouse	Apomys aurorae	x										
Banahaw Forest Mouse	Apomys banahao	x										
Browns' Forest Mouse	Apomys brownorum	x										
Camiguin Forest Mouse	Apomys camiguinensis	x										
Luzon Montane Forest Mouse	Apomys datae	x										
Large Mindoro Forest Mouse	Apomys gracilirostris	x										
Mindanao Mossy Forest Mouse	Apomys hylocoetes	x										
Mindanao Montane Forest Mouse	Apomys insignis	x										
Mount Irid Forest Mouse	Apomys iridensis	x										
Mindanao Lowland Forest Mouse	Apomys littoralis	x										
Lubang Forest Mouse	Apomys lubangensis	x										

English name	Species name	PH	ID	BN	ME	MP	SG	MM	TH	KH	LA	VN
Large Forest Mouse	Apomys magnus	x										
Small Luzon Forest Mouse	Apomys microdon	x										
Mount Mingan Forest Mouse	Apomys minganensis	x										
Least Philippine Forest Mouse	Apomys musculus	x										
Long-nosed Luzon Forest Mouse	Apomys sacobianus	x										
Sierra Madre Forest Mouse	Apomys sierrae	x										
Zambales Forest Mouse	Apomys zambalensis	x										
Isarog Shrew Mouse	Archboldomys luzonensis	x										
Large Cordillera Shrew Mouse	Archboldomys maximus	x										
Kampalili Shrew Mouse	Baletemys kampalili	x										
Lesser Bandicoot Rat	Bandicota bengalensis						x					
Greater Bandicoot Rat	Bandicota indica							x	x	x	x	x
Savile's Bandicoot Rat	Bandicota savilei							x	x	x		x
Large-toothed Hairy-tailed Rat	Batomys dentatus	x										
Luzon Hairy-tailed Rat	Batomys granti	x										
Hamiguitan Hairy-tailed Rat	Batomys hamiguitan	x										
Dinagat Hairy-tailed Rat	Batomys russatus	x										
Mindanao Hairy-tailed Rat	Batomys salomonseni	x										
Mount Isarog Hairy-tailed Rat	Batomys uragon	x										
Berdmore's Rat	Berylmys berdmorei							x	x	x	x	x
Bower's Rat	Berylmys bowersi		x					x	x		x	x
Mackenzie's Rat	Berylmys mackenziei					x		x				x
Manipur Rat	Berylmys manipulus							x				
Large Mindanao Forest Rat	Bullimus bagobus	x										
Carleton's Forest Rat	Bullimus carletoni	x										
Camiguin Forest Rat	Bullimus gamay	x										
Large Luzon Forest Rat	Bullimus luzonicus	x										
Short-footed Luzon Tree Rat	Carpomys melanurus	x										
White-bellied Luzon Tree Rat	Carpomys phaeurus	x										
Fea's Tree Rat	Chiromyscus chiropus							x	x	x	x	x
Langbian Tree Rat	Chiromyscus langbianis							x	x	x	x	x
Thomas's Tree Rat	Chiromyscus thomasi										x	x
Palawan Pencil-tailed Tree Mouse	Chiropodomys calamianensis	x										

English name	Species name	PH	ID	BN	ME	MP	SG	MM	TH	KH	LA	VN
Indomalayan Pencil-tailed Tree Mouse	Chiropodomys gliroides		x			x		x	x	x	x	x
Koopman's Pencil-tailed Tree Mouse	Chiropodomys karlkoopmani		x									
Large Pencil-tailed Tree Mouse	Chiropodomys major		x?	x?	x							
Gray-bellied Pencil-tailed Tree Mouse	Chiropodomys muroides		x		x							
Small Pencil-tailed Tree Mouse	Chiropodomys pusillus		x		x							
Isarog Striped Shrew Rat	Chrotomys gonzalesi	x										
Lowland Striped Shrew Rat	Chrotomys mindorensis	x										
Sibuyan Striped Shrew Rat	Chrotomys sibuyanensis	x										
Blazed Luzon Shrew Rat	Chrotomys silaceus	x										
Luzon Montane Striped Shrew Rat	Chrotomys whiteheadi	x										
Dinagat Bushy-tailed Cloud Rat	Crateromys australis	x										
Panay Bushy-tailed Cloud Rat	Crateromys heaneyi	x										
Ilin Bushy-tailed Cloud Rat	Crateromys paulus	x										
Luzon Bushy-tailed Cloud Rat	Crateromys schadenbergi	x										
Luzon Shrew Mouse	Crunomys fallax	x										
Mindanao Shrew Mouse	Crunomys melanius	x										
Katanglad Shrew Mouse	Crunomys suncoides	x										
Millard's Giant Rat	Dacnomys millardi										x	x
Crump's Soft-furred Rat	Diomys crumpi							x				
Greater Ranee Mouse	Haeromys margarettae		x		x							
Lesser Ranee Mouse	Haeromys pusillus	x	x		x							
Lesser Marmoset Rat	Hapalomys delacouri					x		x	x		x	x
Greater Marmoset Rat	Hapalomys longicaudatus					x		x	x			
Suntsov's Marmoset Rat	Hapalomys suntsovi											x
Javan Bamboo Rat	Kadarsanomys sodyi		x									
Grey Tree Rat	Lenothrix canus		x	x	x	x						
Sundaic Mountain Rat	Leopoldamys ciliatus		x			x						
Diwangkara's Giant Rat	Leopoldamys diwangkarai		x									
Edward's Giant Rat	Leopoldamys edwardsi							x	x		x	x
Herbert's Giant Rat	Leopoldamys herberti							x	x	x	x	x
Miller's Giant Rat	Leopoldamys milleti											x
Neill's Giant Rat	Leopoldamys neilli								x		x	x
Long-tailed Giant Rat	Leopoldamys sabanus		x	x	x	x	x	x	x			x

English name	Species name	PH	ID	BN	ME	MP	SG	MM	TH	KH	LA	VN
Mentawai Long-tailed Giant Rat	Leopoldamys siporanus		x									
Grey-bellied Moss Mouse	Limnomys bryophilus	x										
Long-tailed Moss Mouse	Limnomys sibuanus	x										
Bornean Mountain Maxomys	Maxomys alticola				x							
Small Bornean Maxomys	Maxomys baeodon				x							
Bartels's Javan Maxomys	Maxomys bartelsii		x									
Sumatran Mountain Maxomys	Maxomys hylomyoides		x									
Malayan Mountain Maxomys	Maxomys inas					x						
Broad-nosed Maxomys	Maxomys inflatus		x									
Indochinese Maxomys	Maxomys moi										x	x
Ochraceous-bellied Maxomys	Maxomys ochraceiventer		x		x							
Mentawai Maxomys	Maxomys pagensis		x									
Palawan Maxomys	Maxomys panglima	x										
Rajah Maxomys	Maxomys rajah		x	x	x	x	x					
Red Spiny Maxomys	Maxomys surifer		x	x	x	x	x†	x	x	x	x	x
Tajuddin's Maxomys	Maxomys tajuddinii		x		x							
Whitehead's Maxomys	Maxomys whiteheadi		x	x	x	x			x			
Red-eared Harvest Mouse	Micromys erythrotis										x	x
Pygmy Harvest Mouse	Micromys pygmaeus							x				
Popa Soft-furred Rat	Millardia kathleenae							x				
Ricefield Mouse	Mus caroli							x	x	x	x	x
Fawn-coloured Mouse	Mus cervicolor							x	x	x	x	x
Cook's Mouse	Mus cookii							x	x	x	x	x
Sumatran Shrewlike Mouse	Mus crociduroides		x									
Sheath-tailed Mouse	Mus fragilicauda								x		x	
Little Indian Field Mouse	Mus lepidoides							x				
House Mouse	Mus musculus	x	x	x	x	x	x	x	x	x	x	x
Blyth's Mouse	Mus nitidulus							x				
Indochinese Shrewlike Mouse	Mus pahari							x	x	x	x	x
Shortridge's Mouse	Mus shortridgei							x	x	x	x	x
Volcano Mouse	Mus vulcani		x									
Mount Anacuao Tree Mouse	Musseromys anacuao	x										
Mount Pulag Tree Mouse	Musseromys beneficus	x										

English name	Species name	PH	ID	BN	ME	MP	SG	MM	TH	KH	LA	VN
Mount Banahaw Tree Mouse	*Musseromys gulantang*	x										
Mount Amuyao Tree Mouse	*Musseromys inopinatus*	x										
Brahman Niviventer	*Niviventer brahma*							x				
Bukit Niviventer	*Niviventer bukit*		x			x			x		x	x
Cameron Highlands Niviventer	*Niviventer cameroni*					x						
Confucian Niviventer	*Niviventer confucianus*							x	x			x
Dark-tailed Niviventer	*Niviventer cremoriventer*		x	x	x	x			x			
Smoke-bellied Niviventer	*Niviventer eha*							x				
Montane Sumatran Niviventer	*Niviventer fraternus*		x									
Indomalayan Niviventer	*Niviventer fulvescens*							x	x			x
Limestone Niviventer	*Niviventer hinpoon*								x			
Montane Javan Niviventer	*Niviventer lepturus*		x									
Mekong Niviventer	*Niviventer mekongis*										x	x
Himalayan Niviventer	*Niviventer niviventer*											
Montane Bornean Niviventer	*Niviventer rapit*		x	x	x			x				
Indochinese Mountain Niviventer	*Niviventer tenaster*							x	x	x	x	x
Palawan Mountain Rat	*Palawanomys furvus*	x										
Southern Luzon Giant Cloud Rat	*Phloeomys cumingi*	x										
Northern Luzon Giant Cloud Rat	*Phloeomys pallidus*	x										
Red Tree Rat	*Pithecheir melanurus*		x									
Malayan Woolly Tree Rat	*Pithecheir parvus*					x						
Bornean Tree Rat	*Pithecheirops otion*				x							
Burnished Enggano Rat	*Rattus adustus*		x									
Indochinese Forest Rat	*Rattus andamanensis*							x	x	x	x	x
Ricefield Rat	*Rattus argentiventer*	x	x	x	x	x	x	x	x	x	x	x
Kinabalu Rat	*Rattus baluensis*				x							
Aceh Rat	*Rattus blangorum*		x									
Enggano Island Rat	*Rattus enganus*		x									
Philippine Forest Rat	*Rattus everetti*	x										
Pacific Rat	*Rattus exulans*	x	x	x	x	x	x	x	x	x	x	x
Hoogerwerf's Rat	*Rattus hoogerwerfi*		x									
Sumatran Mountain Rat	*Rattus korinchi*		x									
Lesser Ricefield Rat	*Rattus losea*					x			x	x	x	x

English name	Species name	PH	ID	BN	ME	MP	SG	MM	TH	KH	LA	VN
Mentawai Rat	Rattus lugens		x									
Mindoro Mountain Rat	Rattus mindorensis	x										
White-footed Indochinese Rat	Rattus nitidus							x	x		x	x
Brown Rat	Rattus norvegicus	x	x	x	x	x	x	x	x	x	x	x
Osgood's Rat	Rattus osgoodi											x
Himalayan Rat	Rattus pyctoris											x
House Rat	Rattus rattus	x					x	x	x			
Little Indochinese Rat	Rattus sakeratensis	x	x	x	x	x	x	x	x	x	x	x
Simalur Rat	Rattus simalurensis		x									
Tanezumi Rat	Rattus tanezumi	x	x	x	x	x	x	x	x	x	x	x
Tawi-tawi Forest Rat	Rattus tawitawiensis	x										
Malaysian Wood Rat	Rattus tiomanicus	x	x	x	x	x	x					
Banahao Shrew Rat	Rhynchomys banahao	x										
Isarog Shrew Rat	Rhynchomys isarogensis	x										
Labo Shrew Rat	Rhynchomys labo	x										
Mingan Shrew Rat	Rhynchomys mingan	x										
Northern Luzon Shrew Rat	Rhynchomys soricoides	x										
Zambales Shrew Rat	Rhynchomys tapulao	x										
Lao Limestone Rat	Saxatilomys paulinae										x	x
Kalinga Shrew Mouse	Soricomys kalinga	x										
Leonardo Shrew Mouse	Soricomys leonardocoi	x										
Mountain Shrew Mouse	Soricomys montanus	x										
Sierra Madre Shrew Mouse	Soricomys musseri	x										
Amandale's Rat	Sundamys annandalei						x					
Mountain Giant Rat	Sundamys infraluteus		x	?	x	x						
Bartels's Rat	Sundamys maxi		x		x							
Müller's Rat	Sundamys muelleri	x	x	x	x	x	x	x	x			
Mindanao Dusky Rat	Tarsomys apoensis	x										
Mindanao Spiny Rat	Tarsomys echinatus	x										
Tonkin Limestone Rat	Tonkinomys daovantieni											x
Luzon Short-nosed Rat	Tryphomys adustus	x										
Long-tailed Climbing Mouse	Vandeleuria oleracea							x	x	x	x	x
Vernay's Climbing Mouse	Vernaya fulva							x				x

English name	Species name	PH	ID	BN	ME	MP	SG	MM	TH	KH	LA	VN
Platacanthomyidae												
Vietnamese Pygmy-dormouse	Typhlomys chapensis											×
Sciuridae (Callosciurinae)												
Ear-spot Squirrel	Callosciurus adamsi		×	×	×							
Kinabalu Squirrel	Callosciurus baluensis		×		×							
Grey-bellied Squirrel	Callosciurus caniceps							×	×	×	×	×
Pallas's Squirrel	Callosciurus erythraeus					×		×	×	×	×	×
Variable Squirrel	Callosciurus finlaysonii							×	×		×	×
Hon Khoai Squirrel	Callosciurus honkhoaiensis											×
Inornate Squirrel	Callosciurus inornatus										×	×
Mentawai Squirrel	Callosciurus melanogaster		×									
Sunda Black-banded Squirrel	Callosciurus nigrovittatus		×			×			×			
Plantain Squirrel	Callosciurus notatus		×	×	×	×	×		×			
Bornean Black-banded Squirrel	Callosciurus orestes		×	×	×							
Phayre's Squirrel	Callosciurus phayrei							×	×			
Prevost's Squirrel	Callosciurus prevostii		×	×	×	×		×	×			
Irrawaddy Squirrel	Callosciurus pygerythrus							×				
Stripe-bellied Squirrel	Callosciurus quinquestriatus							×				
Red-throated Squirrel	Dremomys gularis											×
Orange-bellied Squirrel	Dremomys lokriah							×				
Chinese Red-cheeked Squirrel	Dremomys ornatus											×
Perny's Long-nosed Squirrel	Dremomys pernyi							×				×
Red-hipped Squirrel	Dremomys pyrrhomerus											×
Red-cheeked Squirrel	Dremomys rufigenis					×		×	×	×?	×	×
Philippine Pygmy Squirrel	Exilisciurus concinnus	×										
Least Pygmy Squirrel	Exilisciurus exilis		×	×	×	×						
Tufted Pygmy Squirrel	Exilisciurus whiteheadi		×	×	×	×						
Sculptor Squirrel	Glyphotes simus		×			×						
Four-striped Ground Squirrel	Lariscus hosei		×									
Three-striped Ground Squirrel	Lariscus insignis		×	×	×	×			×			
Niobe Ground Squirrel	Lariscus niobe		×									
Mentawai Three-striped Ground Squirrel	Lariscus obscurus		×									
Indochinese Ground Squirrel	Menetes berdmorei							×	×	×	×	×

English name	Species name	PH	ID	BN	ME	MP	SG	MM	TH	KH	LA	VN
Black-eared Squirrel	Nannosciurus melanotis		x	x	x							
Shrew-faced Ground Squirrel	Rhinosciurus laticaudatus		x	x	x	x	x		x			
Sumatran Mountain Squirrel	Sundasciurus altitudinis		x									
Brooke's Squirrel	Sundasciurus brookei		x									
Davao Squirrel	Sundasciurus davensis	x										
Bornean Mountain Ground Squirrel	Sundasciurus everetti		x	x	x							
Fraternal Squirrel	Sundasciurus fraterculus		x									
Horse-tailed Squirrel	Sundasciurus hippurus		x	x	x	x			x			
Busuanga Squirrel	Sundasciurus hoogstraali	x										
Jentink's Squirrel	Sundasciurus jentinki		x	x	x							
Northern Palawan Squirrel	Sundasciurus juvencus	x										
Low's Squirrel	Sundasciurus lowii		x	x	x							
Culion Squirrel	Sundasciurus moellendorffi	x										
Natuna Squirrel	Sundasciurus natunensis		x									
Philippine Squirrel	Sundasciurus philippinensis	x										
Palawan Montane Squirrel	Sundasciurus rabori	x										
Robinson's Squirrel	Sundasciurus robinsoni					x			x			
Samar Squirrel	Sundasciurus samarensis	x										
Southern Palawan Montane Squirrel	Sundasciurus steerii	x										
Malayan Upland Squirrel	Sundasciurus tahan					x						
Slender Squirrel	Sundasciurus tenuis		x	x	x	x	x		x			
Eastern Striped Squirrel	Tamiops maritimus									x	x	x
Western Striped Squirrel	Tamiops mcclellandii							x	x	x	x	x
Cambodian Striped Squirrel	Tamiops rodolphii								x	x	x	x
Swinhoe's Striped Squirrel	Tamiops swinhoei							x				x
Sciuridae (Ratufinae)												
Cream-coloured Giant Squirrel	Ratufa affinis		x	x	x	x	x†		x			
Black Giant Squirrel	Ratufa bicolor		x	x	x	x		x	x	x	x	x
Sciuridae (Sciurinae, Pteromyini)												
Black Flying Squirrel	Aeromys tephromelas		x	x	x	x		x	x			
Thomas's Flying Squirrel	Aeromys thomasi		x	x								
Hairy-footed Flying Squirrel	Belomys pearsonii							x			x	x
Namdapha Flying Squirrel	Biswamoyopterus biswasi							x				x

English name	Species name	PH	ID	BN	ME	MP	SG	MM	TH	KH	LA	VN
Laotian Flying Squirrel	Biswamoyopterus laoensis										x	
Yunnan Woolly Flying Squirrel	Eupetaurus nivamons							?				
Particoloured Flying Squirrel	Hylopetes alboniger							x	x	x	x	x
Bartels's Flying Squirrel	Hylopetes bartelsi		x									
Palawan Flying Squirrel	Hylopetes nigripes	x										
Phayre's Flying Squirrel	Hylopetes phayrei							x	x		x	x
Jentink's Flying Squirrel	Hylopetes platyurus		x		x	x		x	x			
Grey-cheeked Flying Squirrel	Hylopetes sagitta		x									
Sipora Flying Squirrel	Hylopetes sipora		x									
Red-cheeked Flying Squirrel	Hylopetes spadiceus		x	x	x	x	x	x	x	x	x	x
Sumatran Flying Squirrel	Hylopetes winstoni		x									
Horsfield's Flying Squirrel	Iomys horsfieldii		x	x	x	x	x					
Mentawai Flying Squirrel	Iomys sipora		x									
Lesser Pygmy Flying Squirrel	Petaurillus emiliae				x							
Hose's Pygmy Flying Squirrel	Petaurillus hosei			x	x							
Selangor Pygmy Flying Squirrel	Petaurillus kinlochii				x	x						
White-bellied Giant Flying Squirrel	Petaurista albiventer							x	x			
Red-and-White Giant Flying Squirrel	Petaurista alborufus							x	x			
Spotted Giant Flying Squirrel	Petaurista elegans		x	x	x	x		x	x		x	x
Indochinese Giant Flying Squirrel	Petaurista marica							x	x			
Red Giant Flying Squirrel	Petaurista petaurista		x	x	x	x	x†	x	x		x	x
Indian Giant Flying Squirrel	Petaurista philippensis							x	x	x	x	x
Chindwin Giant Flying Squirrel	Petaurista sybilla							x				
Yunnan Giant Flying Squirrel	Petaurista yunanensis							x			x	x
Basilan Flying Squirrel	Petinomys crinitus	x										
Whiskered Flying Squirrel	Petinomys genibarbis		x	x	x	x		x	x			
Hagen's Flying Squirrel	Petinomys hageni		x									
Siberut Flying Squirrel	Petinomys lugens		x									
Mindanao Flying Squirrel	Petinomys mindanensis											
Temminck's Flying Squirrel	Petinomys setosus		x	x	x	x		x	x			
Vordermann's Flying Squirrel	Petinomys vordermanni		x	x	x	x		x	x			
Himalayan Large-eared Flying Squirrel	Priapomys leonardi							x				
Smoky Flying Squirrel	Pteromyscus pulverulentus		x	x	x	x			x			

English name	Species name	PH	ID	BN	ME	MP	SG	MM	TH	KH	LA	VN
Sciuridae (Sciurinae, Sciurini)												
Tufted Ground Squirrel	*Rheithrosciurus macrotis*		x	x	x							
Spalacidae												
Lesser Bamboo Rat	*Cannomys badius*							x	x	x	x	x
Hoary Bamboo Rat	*Rhizomys pruinosus*					x		x	x	x	x	x
Chinese Bamboo Rat	*Rhizomys sinensis*							x				x
Indomalayan Bamboo Rat	*Rhizomys sumatrensis*		x			x		x	x	x	x	x
SCANDENTIA												
Ptilocercidae												
Feather-tailed Treeshrew	*Ptilocercus lowii*		x	x	x	x		x	x			
Tupaiidae												
Bornean Slender-tailed Treeshrew	*Dendrogale melanura*		x?	x?	x							
Northern Slender-tailed Treeshrew	*Dendrogale murina*								x	x	x	x
Northern Treeshrew	*Tupaia belangeri*							x	x	x	x	x
Golden-bellied Treeshrew	*Tupaia chrysogaster*		x									
Banka Island Treeshrew	*Tupaia discolor*		x									
Striped Treeshrew	*Tupaia dorsalis*		x	x	x							
Mindanao Treeshrew	*Tupaia everetti*	x										
Sumatran Treeshrew	*Tupaia ferruginea*		x									
Common Treeshrew	*Tupaia glis*					x	x	x				
Slender Treeshrew	*Tupaia gracilis*		x	x	x							
Mentawai Treeshrew	*Tupaia hypochrysa*		x	x								
Horsefield's Treeshrew	*Tupaia javanica*		x									
Long-footed Treeshrew	*Tupaia longipes*		x	x	x							
Lesser Treeshrew	*Tupaia minor*		x	x	x	x			x			
Mountain Treeshrew	*Tupaia montana*		x	x	x							
Palawan Treeshrew	*Tupaia palawanensis*	x										
Painted Treeshrew	*Tupaia picta*		x	x	x							
Kalimantan Treeshrew	*Tupaia salatana*		x									
Ruddy Treeshrew	*Tupaia splendidula*		x									
Large Treeshrew	*Tupaia tana*		x	x	x	x						

MARINE MAMMALS

English name	Species name	IUCN
CETACEA		
Balaenopteridae		
Common Minke Whale	Balaenoptera acutorostrata	LC
Sei Whale	Balaenoptera borealis	EN
Bryde's Whale	Balaenoptera brydei	LC
Eden's Whale	Balaenoptera edeni	DD
Blue Whale	Balaenoptera musculus	EN
Omura's Whale	Balaenoptera omurai	DD
Fin Whale	Balaenoptera physalus	VU
Humpback Whale	Megaptera novaeangliae	LC
Delphinidae		
Common Dolphin	Delphinus delphis	LC
Pygmy Killer Whale	Feresa attenuata	LC
Short-finned Pilot Whale	Globicephala macrorhynchus	LC
Risso's Dolphin	Grampus griseus	LC
Fraser's Dolphin	Lagenodelphis hosei	LC
Irrawaddy Dolphin	Orcaella brevirostris	EN
Killer Whale	Orcinus orca	DD
Melon-headed Whale	Peponocephala electra	LC
False Killer Whale	Pseudorca crassidens	NT
Indo-Pacific Humpbacked Dolphin	Sousa chinensis	VU
Australian Humpback Dolphin	Sousa sahulensis	VU
Pantropical Spotted Dolphin	Stenella attenuata	LC
Striped Dolphin	Stenella coeruleoalba	LC
Spinner Dolphin	Stenella longirostris	LC
Rough-toothed Dolphin	Steno bredanensis	LC
Indo-pacific Bottlenose Dolphin	Tursiops aduncus	NT
Common Bottlenose Dolphin	Tursiops truncatus	LC
Kogiidae		
Pygmy Sperm Whale	Kogia breviceps	LC
Dwarf Sperm Whale	Kogia sima	LC
Phocoenidae		
Indopacific Finless Porpoise	Neophocaena phocaenoides	VU
Physeteridae		
Sperm Whale	Physeter macrocephalus	VU
Ziphiidae		
Longman's Beaked Whale	Indopacetus pacificus	LC
Blainville's Beaked Whale	Mesoplodon densirostris	LC
Ginko-toothed Beaked Whale	Mesoplodon ginkgodens	DD
Cuvier's Beaked Whale	Ziphius cavirostris	LC
SIRENIA		
Dugongidae		
Dugong	Dugong dugon	VU

FURTHER INFORMATION

A number of books and online resources have been extremely useful in the development of this book. Additionally, there are numerous peer-reviewed publications that are highly informative, such as Duckworth and Pine's 'English names for a world list of mammals, exemplified by species of Indochina' (*Mammal Review*, 2003, Vol. 33, 151–173), which guided the further improvement and standardisation of common English names for Southeast Asian mammals. We recommend the following, but there are also many other useful resources that are not listed here. Also included are the specialist groups, which are networks of experts in the respective species groups, under the umbrella of the IUCN's Species Survival Commission.

REFERENCES AND FURTHER READING

Baker, N. and Lim, K. (2008). *Wild Animals of Singapore: a Photographic Guide to Mammals, Reptiles, Amphibians and Freshwater Fish.* Draco Publishing & Distribution and Nature Society Singapore, Singapore.

Carwardine, M. (2006). *Whales, Dolphins and Porpoises.* HarperCollins, London, UK.

Ecology Asia (2012). www.ecologyasia.com. Singapore.

Francis, C. M. (2001). *A Photographic Guide to the Mammals of Southeast Asia.* New Holland, London, UK.

Francis, C. M. (2019). *Field Guide to the Mammals of South-East Asia*, 2nd edition. Bloomsbury, London, UK.

Groves, C. and Grubb, P. (2011). *Ungulate Taxonomy.* Johns Hopkins University Press, Baltimore, MD, USA.

IUCN (2012). *The IUCN Red List of Threatened Species. Version 2012.1.* www.iucnredlist.org.

Lekagul, B. and McNeely, L. (1977). *Mammals of Thailand.* Association for the Conservation of Wildlife, Bangkok, Thailand.

Lim, N. (2007). *Colugo: the Flying Lemur of South-east Asia.* Draco Publishing & Distribution and the National University of Singapore, Singapore.

Kingston, T., Lim, B. L. and Zubaid, A. (2006). *Bats of Krau Wildlife Reserve.* Penerbit Universiti Kebangsaan Malaysia, Bangi, Malaysia.

Krëb, D. (2004). *Facultative river dolphins: conservation and social ecology of freshwater and coastal Irrawaddy Dolphins in Indonesia.* Institute for Biodiversity and Ecosystem Dynamics / Zoologisch Museum Amsterdam.

Kruuk, H. (2006). *Otters: Ecology, Behavior and Conservation.* Oxford University Press, New York, NY, USA.

Parr, W. K. J. and Hoang Xuan Thuy (2008). *A Field Guide to the Large Mammals of Vietnam.* People and Nature Reconciliation (PanNature). Thong Tan Publishing House, Hanoi, Vietnam.

Payne, J., Francis, C. M. and Phillipps, K. (1985). *A Field Guide to the Mammals of Borneo.* The Sabah Society, Malaysia.

Redmond, I. (2008). *Primates of the World.* New Holland, London, UK.

Smith, A. T. and Xie, Y. (2008). *Mammals of China.* Princeton University Press, Princeton, NJ, USA.

Synopsis of Philippine Mammals (2012). archive.fieldmuseum.org/philippine_mammals.

Wilson, D. E., Mittermeier, R. A., Lacher, T. E. Eds (2019) *Handbook of the Mammals of the World*, vols 1–9. Lynx Edicions, Barcelona, Spain.

RELEVANT IUCN SSC SPECIALIST GROUPS

Asian Elephant Specialist Group – www.asesg.org
Asian Rhino Specialist Group – www.iucn.org/our-union/commissions/group/iucn-ssc-asian-rhino-
    specialist-group
Asian Wild Cattle Specialist Group – www.asianwildcattle.org
Bat Specialist Group – www.iucnbsg.org
Bear Specialist Group – globalbearconservation.org
Canid Specialist Group – www.canids.org
Caprinae Specialist Group – iucncaprinaesg.weebly.com
Cat Specialist Group – www.catsg.org
Cetacean Specialist Group – www.iucn-csg.org
Deer Specialist Group – www.deerspecialistgroup.org
Lagomorph Specialist Group – iucn.org/our-union/commissions/group/iucn-ssc-lagomorph-
    specialist-group
Otter Specialist Group – www.otterspecialistgroup.org
Pangolin Specialist Group – www.pangolinsg.org
Primate Specialist Group – www.primate-sg.org
Sirenia Specialist Group – iucn.org/our-union/commissions/group/iucn-ssc-sirenia-specialist-group
Small Carnivore Specialist Group – smallcarnivore.org
Small Mammal Specialist Group – small-mammals.org
Tapir Specialist Group – www.tapirs.org
Wild Pig Specialist Group – www.iucn-wpsg.org

ACKNOWLEDGEMENTS

This book would not have been possible without input and support from many individuals working on mammal conservation in Southeast Asia, and the very generous contributions of photographs from these same individuals, as well as others.

All photographers are credited on page 176. However, we would especially like to thank the following for helping track down hard-to-come-by photos, for providing information and for being very supportive of this project. We are fortunate to call them friends. Thanks to Abraham Mathew, Barney Long, Celine Low, Danielle Krëb, Jasmine Steed, James Eaton, Matt Linkie, Matt Struebig, Neil Furey, Nick Baker, Pilar Salajeno, Resit Sozer, Sabine Schoppe and Serge Wich. We extend a special thanks to Will Duckworth for providing extremely useful information and advice, for reviewing earlier drafts, and for tirelessly responding to countless questions!

We would also like to thank John Beaufoy, Ken Scriven, Rosemary Wilkinson, Hugh Brazier and David Price-Goodfellow for pulling the whole project together.

A note of mention to the researchers and conservation workers, who tirelessly toil in the field to better understand and protect the mammals of Southeast Asia.

Photo credits

Photos are denoted by a page number followed where relevant by t (top), b (bottom), l (left) or r (right.)

Andie Ang: 41. **Nick Baker, EcologyAsia.com:** 16, 17l, 17r, 20r, 21, 22r, 26, 37, 61, 89, 118, 119, 121, 123, 124l, 124r, 133, 134, 135. **Elizabeth A. Burgess:** 93l, 93r. **Dan Challender/ Carnivore and Pangolin Conservation Program, Vietnam:** 12l. **Iing Cikananga:** 100l, 100r. **Leanne Clark/Carnivore and Pangolin Conservation Program, Vietnam:** 13. **Roger G. Dolorosa/Western Philippines University:** 104, 138. **Vilma D'Rozario/Cicada Tree Eco-Place:** 70, 86. **Duc Hoang Minh:** 46. **Nicole Duplaix:** 64, 65l, 65r, 66. **James Eaton/Birdtour Asia:** 12r, 19, 38, 43, 44l, 44r, 54, 55t, 69, 75, 94, 108, 126, 129l, 129r, 131. **Fletcher & Baylis:** 56, 68, 106, 107l, 107r, 111, 136l, 136r, 137. **Charles M. Francis** 20l, 22l, 23, 25, 27, 28, 29, 79t. **Gabriella Fredriksson:** 39. **Neil Furey:** 24. **Stephen Hogg/Wildtrack Photography:** 4, 7, 14b, 51t, 52t, 52b, 55bl, 55br, 80l, 80r, 95t, 95b, 97l, 97r. **Kadoorie Farm & Botanic Garden, Hong Kong:** 78. **Kae Kawanishi/MYCAT:** 77, 81, 84, 96, 99, 115. **Kimabajo:** 15, 58r, 101, 103, 105l. **Danielle Krëb/ Conservation Foundation for Rare Aquatic Species of Indonesia:** 88, 90, 91, 92. **Ch'ien C. Lee/Wild Borneo Photography:** 30, 73. **Celine Low:** 76, 113, 125, 127, 128. **Joey Markx:** 34. **Abraham Mathew:** 112l, 112r. **Abraham Mathew/Singapore Zoo and Night Safari:** 45, 57, 87l, 87r. **Anuar McAfee:** 51b. **Mohamed & Wilting, Sabah Wildlife Department, Sabah Forestry Department:** 14t, 63, 74, 79b, 82, 83, 85. **Richard Moore, International Animal Rescue:** 31, 33. **NHPA/Gerald Cubitt:** 72, 139. **NHPA/Oscar Dominguez:** 102. **NHPA/Daniel Heuclin:** 114. **Pilar Saldajeno:** 109r. **Chris R. Shepherd:** 9, 36, 42, 47, 48l, 48r, 49l, 49tr, 49br, 50l, 50r, 59, 71l, 71r, 105r, 120. **Coke & Som Smith/www.cokesmithphototravel.com:** 40, 46. **John Steed:** 122. **Sabine Stolzenburg:** 67. **Ulrike Streicher:** 32, 62. **Rob Tizard:** 18. **Jonah van Beijnen, Centre for Sustainability, Philippines:** 130l, 130r. **WCS Myanmar Program:** 60b, 60t, 110, 116. **Serge Wich:** 35. **Peter Widmann:** 109l, 132. **Christy Williams:** 53l, 53r. **Wong Siew Te/Bornean Sun Bear Conservation Centre:** 58l. **WWF/CTNPCP/Mike Baltzer:** 98. **WWF/Toon Fey:** 117.